STUDIES IN ENVIRONMENTAL POLLUTION

The estimation of pollution damage

STUDIES IN ENVIRONMENTAL POLLUTION

The geography of pollution   C. M. Wood *et al.*

Town planning and pollution control   C. M. Wood

P. J. W. SAUNDERS

The estimation of pollution damage

Manchester University Press

Published by
MANCHESTER UNIVERSITY PRESS
Oxford Road, Manchester M13 9PL

UK ISBN 0 7190 0629 5

TD
/77
.S23

Printed in Great Britain by
Eyre and Spottiswoode Ltd at Grosvenor Press, Portsmouth

CONTENTS

LIST OF TABLES

LIST OF FIGURES

General introduction

It is apparent that the recent growth in public concern over environmental pollution has not been matched by a corresponding increase in understanding about the form it takes, the sources from which it originates, its effects or the manner in which it is (or can be) controlled. In part this is a problem of communication and in part a problem of research technique and objective. Confronted by a large, uncharted territory, it is tempting to retreat into some small corner, applying the tools of one's discipline to an aspect of one environmental medium's problem.

However, waste creation and disposal, whether in gaseous, liquid, solid or noise form, are interrelated problems. Similarly, the significance of scientific research into the effects of pollution or the technical methods of controlling it can be established only with reference to the type of society in which the pollution phenomenon occurs. Equally, the examination of appropriate pollution control levels and the administrative instruments by which these are achieved is a hollow exercise unless the necessary scientific backing is to hand. In short, a genuinely interdisciplinary approach to the whole problem of waste is ideally required.

It is considerations of this kind which have led to the introduction of this series and which have conditioned its form. It is intended for three types of reader. First, the academic specialist who is anxious to gain a wider understanding of environmental pollution, particularly from the standpoint of disciplines other than his own. Second, those who are engaged in a professional capacity, technical or administrative, in some aspect of pollution control. Finally, the more general reader, who, we hope, will be attracted by the thoroughness of treatment, achieved with the minimum of technical terminology.

There is, in the series, a deliberate intention to treat the subject matter in an objective manner but, in so doing, not to avoid contentious issues. Apart from the existence of conflicting evidence and interpretation on some important issues, there are also differences of view on the research methods by which knowledge will be advanced. These are important matters which require careful examination.

Finally, it should be clear that though the monographs in this series share the same broad purpose, the particular treatment in each is the responsibility of the author and is not necessarily accepted by other contributors to the series.

Norman Lee

PREFACE

Pollution of the air, fresh water, the sea and the land has diverse effects upon the environment. Each of these effects results ultimately in damage to man and his environment. The detection, measurement and estimation of damage are important stages in the rational control of pollution and in the determination of priorities for future pollution research.

There are few books dealing solely with pollution damage, and most of them specialise in a relatively narrow range of pollution problems. The purpose of this book is twofold. First, it provides an outline survey of the current state of knowledge on the physical damage caused by the main forms of pollution. Then it examines, in greater detail, the basic common principles and problems of damage estimation in order to indicate the strategies and information needed to achieve sounder estimates of pollution damage.

The book is primarily intended for university and college students taking courses in environmental pollution and for those professionally engaged in pollution control, although it should also be useful to researchers examining the effects of pollution. Since the technical terminology is kept to a minimum, it should also interest the more general reader with a concern for environmental problems.

Although the views expressed are my own, I gratefully acknowledge the ideas, factual material and valuable criticism contributed by my former colleagues of the Pollution Research Unit, especially Dr Norman Lee, Mr Chris Wood and Dr Edward Bellinger. My thanks are due also to Professor D. H. Valentine for his advice and comments, to Miss Cathryn Sharples, who prepared the manuscript, and to my wife for her tolerance and encouragement throughout the evolution of the book.

P. J. W. Saunders

CHAPTER 1

Pollution damage: an introduction

The effects of pollution on man and the environment are the subject of much discussion[1] and research. Obvious adverse phenomena such as the effects of sewage effluent upon fisheries still cannot be quantified as physical, social or economic losses. This is true especially of subtle effects of chronic pollution by low concentrations of environmental contaminants. Much uncertainty arises from the paucity and unreliability of data relating to both polluted and unpolluted environments coupled with the complex natures of the relationships between waste generation, pollution in different media (e.g. air, water) and the effects of pollution upon the environment as a whole. Furthermore few objective attempts have been made to assess trends in pollution and damage or to identify the margins of error inherent in such work. A comprehensive approach to the pollution issue in which the methods of relevant disciplines are integrated offers a way of understanding the pollution problem.

Pollution causes changes in the quality and quantity of sensitive receptors whose functions as resources are also changed. The evaluation of such damage is an important but difficult process in the rational allocation of scarce financial resources to control pollution. Changes in value are irrelevant, however, to the technical estimation of physical damage; a pollutant will cause the same physical damage to sensitive receptors exposed under identical environmental conditions both now and in the future. Reliable estimates of physical damage are essential for the determination of priorities in pollution control. This book should be viewed in this context.

WHAT ARE POLLUTION AND POLLUTION DAMAGE?

For present purposes pollution is defined as 'The introduction by man of waste matter or surplus energy into the environment which directly or indirectly causes damage to man and his environment *other than* himself, his household, those in his employment or those with whom he has a direct trading relationship'.[2]

Pollution is confined to waste disposals having adverse effects upon 'third parties'. Although the importance of self-inflicted (e.g. cigarette smoking), domestic and occupational (e.g. inhalation of dust at work) exposures is recognised, they are excluded here. However, it is sometimes necessary to take account of such exposures when estimating damage due to environmental pollution. For instance, bronchitis in man may be induced by a mixture of cigarette smoking, occupational exposure to smoke and local environmental smoke pollution from coal fires.

Wastes and energy include gases, particulates, liquids, solids, noise, heat, radiations and vibrations discharged into the atmosphere, fresh waters, ground waters, estuaries and seas, and on to land. The definition excludes identical materials released without human involvement by natural processes (e.g. sulphur dioxide from volcanoes). A distinction is drawn between wastes and pollutants. Wastes are not pollutants until they are discharged to and cause damage in the environment. Furthermore, wastes emitted at one point may be transmitted through various media (e.g. air, water) to produce pollutants at different locations and points in time and in different concentrations, often with alterations to their chemical and physical characteristics.

'Damage' includes all adverse effects on man, his artefacts and the environment. It thus embraces a range of different concepts. *Acute* damage refers to the effects of short-term exposure to relatively high concentrations of a pollutant. The response of the receptor is usually instantaneous and damage is permanent. Complementary terms include *acute* exposure and *acute* response, etc. *Chronic* damage results from exposure to lower concentrations of a pollutant over much longer periods of time. The damage may be temporary or permanent, depending on the definition adopted.The transitions from acute to chronic and from permanent to temporary damage are often difficult to determine. This is reflected in the identification of damage symptoms. Under conditions of greatly prolonged exposure to very low concentrations of a pollutant, symptoms are extremely difficult to detect because they are obscured by similar responses to other environmental stimuli. Responses of this type have been called *subtle, cryptic* and *sub-chronic* damage. Classic examples include changes in animal behaviour and loss of plant yield without obvious symptoms of damage.[3, 4, 5]

With some pollutants there may be a *latent* period between exposure and the appearance of damage. This is especially characteristic of carcinogenic, radioactive and chemical pollutants, which may have latent periods of thirty years or longer in relation to certain forms of cancer. A different type of latency occurs in the temporal delays in the diffusion of persistent pollutants through the environment and along food chains, causing damage to receptors remote from the original point of discharge.

It is common to equate injury (effect of pollution) with damage (loss or consequence of pollution). Injury may be defined as any adverse physical response to, or direct effect of, pollution, but it is often also used to describe symptoms. Strictly speaking, physical damage should be limited to the result or consequence of that response (e.g. loss of productivity or quality). It is this physical damage that impinges directly upon the desired use of the receptor by man,[6] whether economic or social (e.g. aesthetic). Socio-economic values are implicit in assumptions of aesthetic, social and ecological worth[7]. Damage is considered to include, therefore, all physical losses incurred by a receptor and the environment which result in reductions in value of social and economic assets. (This does not assume that economic values may be assigned to all forms of damage.)

The relationship between injury and damage is not constant. For instance, although severe injury of the foliage of root crops can be caused by air pollutants, yield losses are negligible if exposure occurs just prior to harvest. Earlier exposures may increase yield losses substantially despite changes in sensitivity of foliage to air pollution (table 1). Conversely, tainting of fish or slight injury to salad crops

Table 1 Effect of sulphur dioxide upon injury and yield at different stages in the growth of radish

Leaf stage	*Leaf area injured (%)*	*% yield (weight of root) loss compared with unexposed plants*
Emergence	90	90
2	37	30
4	12	5
4–6	26	40
6	82	60
Mature	86	0

Plants exposed to 600–800 μg SO_2 for twenty-four to forty-eight hours after growing to indicated leaf stage (number of leaves fully emerged at the time of fumigation).[8]

(e.g. lettuce) causes little or no physical loss of productivity and yet the products are unmarketable; here the economic damage is much greater than any physical loss. Much more complex situations arise where the injury to a receptor can result in different losses depending on the ownership and function of the receptor. For example, air pollution injury to an ornamental tree has quite different social and economic consequences for the nursery grower of the tree and for the householder or landowner.

It is not sufficient to consider the *obvious* response of a receptor to pollution when appraising damage. For example, the exacerbation of chronic bronchitis by air pollution has secondary consequences in loss of mobility, incapacity for work and early mortality in the same receptor (figure 1). The reactions (e.g. pain and suffering) of relatives are indirect yet important responses to these consequences. Similarly, indirect damage may occur in a food chain in response to direct damage to lower trophic orders. Ideally the boundaries of damage studies would be determined by the points at which these chain reactions become insignificant.

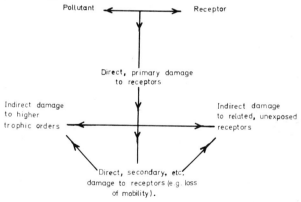

FIGURE 1   Types of damage

Complications may arise where social responses are used as direct indicators of damage. For instance, numbers of complaints by the public to controlling authorities may be used for such purposes, but they suffer from limitations such as a strong tendency to be orientated towards specific sources of pollution (e.g. noise from factories) or to understate the impact of pollution upon the lower-income groups of the population (chapter 6). Generally, complaints are most suitable as indicators of short-term pollution incidents from readily identified sources.

Damages to materials (e.g. buildings) and amenity (e.g. recreational facilities)[9,10] are probably the best examples of responses to pollution in which the damage may in principle be measured directly as an economic loss. In the former case it is possible to measure physical damage in terms of loss of strength, colour fading, etc, but damage is expressed more obviously as increased costs of maintenance (e.g. cleansing), replacement of parts and over-specification (e.g. protective coatings). Damage to amenity, on the other hand (chapter 6), is associated most readily with reductions in consumer appreciation and usage as reflected in visit rates and expenditure on a particular recreation (e.g. angling) in a polluted area. However, in practice the economic estimation of pollution damage is fraught with many methodological and data problems which are not the primary subject for discussion here.[9,11]

DAMAGE AND POLLUTION CONTROL

Figures 2 and 3 illustrate the wastes that can be generated within an economic system and the factors influencing waste output in any given production process.[10,11] Pollution damage, a component of these systems, may be minimised by several alternative control procedures.[10,11] Thus an industry can choose one or more of the following options: cease production, change the composition of products to the least polluting mixture, increase the ratio of product to waste (i.e. greater efficiency of production), increase the re-use and recycling of wastes, improve waste treatment and disposal, and protect sensitive receptors (e.g. sound insulation). Ideally the technical efficiency of waste disposal and pollution control would rely upon the cheapest of these options in relation to the capacity of the local environment to absorb wastes without any damage. In reality it is in society's interests to minimise such costs in relation to the damage caused by pollution and the economic benefits achieved by its control.

Various systems of control have been proposed, involving different charges, taxes, etc, upon polluters and society.[9-13] An important part of most systems is the economic damage function, i.e. the cost of damage in relation to the level of pollution (figure 4). This is derived by evaluation of the relationship between the level of pollution and the degree of physical damage. It may be used to calculate the economic benefits of incremental reductions in pollution achieved at different stages of control. Unfortunately it is in these areas that data are most deficient. This book is devoted, therefore, to the fundamental aspects of physical pollution damage and of its measurement that are directly relevant to the economic evaluation of pollution damage and its control.

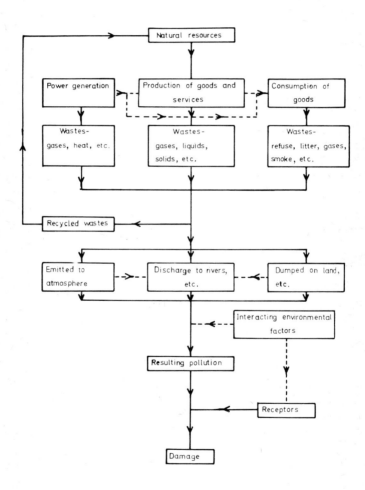

FIGURE 2   Waste creation within an economic system [9-12]

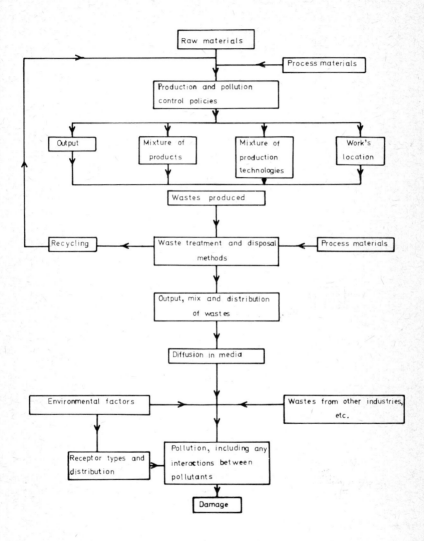

FIGURE 3 Pollution damage and its determinants in a production process of an individual industry[9-12]

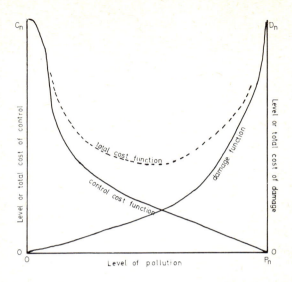

FIGURE 4   The role of the damage function in optimal pollution control.[9-12] All
functions are hypothetical

INTRODUCTION TO CHAPTERS 2–7

The next chapter (chapter 2) reviews pollution and damage with reference to this country. A further chapter (chapter 3) is devoted to global pollution. These both serve as background material for the rest of the book.

The basic methods of damage measurements are discussed in chapter 4. This is followed by more detailed examinations of the relationship between pollution and damage, and the nature of responses to pollution (chapter 5). Chapter 6 discusses the broad strategy of damage studies, including variations in practical and desk-study techniques. A framework for the analysis of damage is proposed and some fundamental problems of damage estimation are examined. The final chapter (chapter 7) reviews the basic points of the book and briefly indicates priority areas for damage research.

CHAPTER 2

A brief survey of pollution and damage

This survey briefly reviews the major types of pollution.[13] Four points should be borne in mind:

1   Any pollutant in sufficient quantities or concentrations causes damage; its importance results from the frequency and distribution of damaging concentrations, not from its mere presence in the environment.

2   Similarly, rising trends in a pollutant do not automatically result in damage if the concentrations are insufficient to cause injury.

3   The isolated occurrence of severe damage must not be confused with the probability of its widespread occurrence.

4   The few estimates of damage, costs of damage and receptor values presented in various sections should be treated with considerable caution.

Acute damage is confined mostly to limited areas around industrial and accident (e.g. tanker crash) sites. Acute water pollution is quite common, however, in the lower reaches of rivers and in estuaries (e.g. of the Mersey). Given adequate investment in control technology and enforcement, prevention of acute pollution is relatively easy.

Chronic pollution and damage are probably more common. Detection and control are more difficult, however, because there is often a latent period between exposure and response. The latter is frequently non-specific (i.e. the response is induced by several pollutants and environmental stresses). At even lower levels of exposure damage takes more subtle or cryptic forms (e.g. behavioural changes in animals). These are difficult to detect without expert investigation. Furthermore such pollution usually originates from many widespread sources, which presents formidable control problems (e.g. control of vehicle emissions).

When comments are made about future trends in pollution and damage it must be appreciated that these are derived by extrapolation of past trends and current situations, which demands the existence of time-series data (e.g. annual average pollution levels over the last ten years or so).[10] In many cases these are sadly lacking. There are plans, however, to remedy the situation on a national scale.[14]

AIR POLLUTION

Impact
Air pollution affects humans, animals, plants and materials, and contaminates soils and surface waters, including the marine environment.[10,13,15,16] There are detectable effects upon local meteorological conditions and longer-range impacts upon other countries and global climate (chapter 3).

Generally, sulphur dioxide (SO_2) with sulphur acids and particulates (e.g. dust, grit, soot, tar, aerosols), are regarded as the most ubiquitous atmospheric pollutants with the greatest overall impact. In urban/industrial areas positive correlations can be established between the levels of these pollutants and human mortality and morbidity, particularly those due to bronchitis and emphysema.[16] Short-term changes in death rates and discomfort (e.g. amongst bronchitics) can be detected where pollution levels exceed about 500 μg smoke and 250 μg SO_2/m^3 air per twenty-four hours for a few days. Under current conditions, however, response is usually associated with advanced mortality and long-term morbidity due to chronic exposures of between 150 and 250 μg smoke and SO_2 during winter months. (The combined action of SO_2 and smoke is probably greater than the effect of either pollutant alone.) Associations between these pollutants and other diseases such as lung and other cancers, asthma, ischaemic and hypertensive disorders, etc, have been claimed, but with far less certainty. However, these chronic responses are nonspecific, being associated with habits (e.g. smoking), occupational exposure, socio-economic status, population density, age, sex, regional/ethnic groupings, quality of water supplies, climate and other factors, as well as with air pollution.[17] There is great variation in estimates of impact of SO_2 and smoke upon human health. Recent estimates [18,19] range between 20 and 50 per cent of mortality and morbidity due to bronchitis and emphysema, with smaller values for other causes of death. The annual costs associated with such damage have been tentatively estimated at between £21 million and £63 million in the UK. [18,19]

The damage to human health caused by other atmospheric pollutants is less certain.[16] The impact of vehicle emissions is uncertain because environmental concentrations have not been effectively monitored. There are also problems in determining the most important pollutant intake in certain cases (e.g. lead in air or food or water[20]). Furthermore responses to most pollutants are non-specific and exceedingly subtle. Thus reports of carbon monoxide (CO) contributing to morbidity, especially due to heart diseases, and to more subtle effects such as loss of co-ordination, are difficult to confirm under environmental conditions. Urban lead levels are generally higher than rural concentrations, particularly close to certain industries and busy roadways.[20] The latter suggest that vehicle emissions may be important sources of local lead pollution.[21,22] This applies also to lead poisoning. The total intake in urban man is sufficient to maintain average levels of between 20 and 30 μg Pb/l blood. Considering the closeness of this average to levels (e.g. ≥ 36 μg Pb/l blood) regarded as unsafe for children, the situation cannot be regarded as satisfactory. Apart from isolated instances, other industrial and urban pollutants such as fluorides, heavy metals, hydrocarbons and airborne radio-nuclides cannot be linked definitely with damage to human[23] and animal health at current levels.

Apart from its aesthetic disadvantages, smoke pollution is reported to affect weather conditions (e.g. sunlight penetration),[24] especially during inversion periods and sometimes in conjunction with aerosol formation (e.g. Teesside mist). Visibility in many areas of the United States and in Japan has been seriously limited by oxidant pollution derived by photo-oxidation of unburnt hydrocarbons under inversion conditions in bright sunlight. Oxidant pollutants have been detected here.[25] Because of the success of the Clean Air Acts sunlight intensity is increasing in many urban areas. Serious oxidant smog problems are not anticipated, but the situation deserves close observation, particularly in view of some oxidant damage to plants reported by university research workers.

The air also serves as a medium for odours and noises, both of which can cause degradation of local residential amenity around some industries. Loss of efficiency at work and general psychological stress[26] due to noise are suspected. Fortunately noise from road and air transport, commerce, entertainment, industry and domestic sources rarely reaches levels ($\geq$90 dBA) in the external environment sufficient to cause pain and permanent loss of hearing. Nevertheless more than 65–70 dBA is considered annoying or irritating—a level regularly exceeded close to airports, factories and busy roadways. In the last case approximately 30 per cent of the urban population is exposed to such levels for about 10 per cent of the day (peak traffic periods).[27]

Damage to plant life by air pollution is difficult to detect because of the non-specific nature of injury symptoms.[28,29] Furthermore, because pollution is seldom a new phenomenon, damage has been accrued over many months or years; subtle changes favouring less sensitive plants growing in permanent communities are suspected.

Acute damage due to particulate (e.g. acidic aerosols, alkaline dusts) and gaseous (e.g. SO_2, NO_x, ethylene, peroxyacyl nitrates) pollutants is mostly localised around specific industrial activities (e.g. metal smelting, sulphuric acid production). For instance, ethylene from industrial sources[29] is known to cause substantial injury in the form of leaf scorch and flower drop; severe yield losses have been observed in sensitive species such as potato.

Chronic overt damage is probably more widespread. Certainly sensitive organisms such as lichens, bryophytes and fungi are limited by quite low concentrations of SO_2 (e.g. 30–100 μg SO_2/m^3 per twenty-four hours.)[30,31] Under experimental conditions prolonged exposure to similar levels (90–150 μg SO_2) can induce reductions in the productivity of sensitive plant species (e.g. grasses[5]). Photochemical pollutants derived from vehicular and industrial emissions can injure plants,[29] although there has only recently been a positive report of such injury to sensitive indicator plants (e.g. Bel—W3 tobacco) in this country.

Particulate contaminants (e.g. cement dust, soot, fluorides, phosphate dusts) accumulate upon foliage, blocking stomata and altering metabolic activity, often without causing visible injury. They also reduce the amount of light available for photosynthesis.[29] Surface deposits of persistent particulates (e.g. heavy metals, pesticides, phosphates, fluorides, radio-nuclides) are potentially more important because they can cause indirect damage to herbivorous wildlife and to stock animals (e.g. fluorosis[32]) by contaminating vegetation used as feedstuff. In addition they contribute to man's pollutant uptake (e.g. metals and pesticides on crops, radio-nuclides in milk). Obvious damage by these types of pollution is exceptional and localised.[33] Available data do not suggest any cause for serious concern about the long-term consequences of chronic contamination of foodstuffs under present conditions.[20]

Atmospheric gases (especially SO_2), acidic particulates and salt aerosols accelerate the corrosion and deterioration of materials, including stonework, metals, wood, fabrics, clothing, dyes and paint-work, thereby reducing their longevity and value and reducing the general quality of the human environment. Air pollution contributes to replacement and maintenance costs[19, 29] of materials. The presence of surface moisture, and high temperatures and relative humidities with SO_2 greatly accelerate metal corrosion. Soiling (viz. surface con-tamination) by particulates aggravates the corrosion process and directly degrades local visual amenity. The latter alone results in increased cleansing, which accelerates wear, substantially increasing overall maintenance and replacement costs. Total costs of air pollu-tion damage to materials in the UK have been estimated at £80 million[19] but the precise figure is uncertain because calculations are complicated by over-specification, the substitution of less sensitive materials and the use of protective coatings (e.g. zinc galvanising, painting). These practices appear to reduce replacement and maintenance requirements but increase the total costs ultimately attributable to corrosion. 'Habitual' maintenance programmes (e.g. the weekly wash) also obscure true maintenance costs in the polluted areas.

Acute damage to soils exposed to acidic gases and corrosive fall-out is very localised,[28] mostly in industrial areas with a history of uncontrolled waste disposal. Characteristically these soils are denuded of vegetation, sterilised and subject to erosion. The longer-term chronic effects of air pollution upon soils are obscure (chapter 3). Some leaching of nutrients and acidification of some lighter soils, in polluted areas, probably occur. These may be corrected readily by adequate fertiliser and lime applications. On the other hand, incipient sulphur deficiency, due possibly to a combination of modern sulphur-free fertilisers and a decline in atmospheric sulphur dioxide concentrations, has been observed in upland pastures.[34]

Fall-out of persistent pollutants (e.g. heavy metals) can result in soil contamination. These pollutants may be taken up by plants, resulting in contaminated forage for herbivorous animals and ulti-mately contributing to the contamination of human food. Again, acute cases are limited to the vicinity of certain industrial activities. Levels of some metals (e.g. lead) in urban soils are higher than those in rural areas, suggesting extensive chronic pollution.[35] However, plants from soils in these areas form a very small part of human diet and contribute little to total pollutant intake.

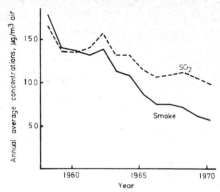

FIGURE 5   Trends in urban UK smoke and sulphur dioxide ground-level concentrations

Trends

Over the last decade there has been a decline in emissions and ground-level concentrations of particulates (figure 5).[17] These have been reflected in improvements in visibility,[24] decreases in melanic variants of certain moth populations,[36] improved success of planting programmes[33] and generally cleaner conditions in urban areas. Emissions of sulphur dioxide have increased over the last decade but, because of a switch from low-level (e.g. domestic chimneys) to high-level (e.g. industrial and power-station chimneys) in the bulk of emissions, ground-level concentrations have declined (figure 5).[2,15] There is circumstantial evidence that other forms of plant damage and metal corrosion due to SO_2 pollution have declined.

The most dramatic result of these reductions in ground-level concentrations, however, has been an improvement in the health of urban man.[16] Acute winter episodes of pollution are now rare, thus greatly reducing incidences of high mortality amongst bronchitic patients. Furthermore it is becoming increasingly difficult to correlate pollution levels with rates of urban mortality and morbidity due specifically to bronchitis and emphysema;[17,37] other factors associated with mortality have become more predominant. With local exceptions, it is anticipated that the downward trends in smoke pollution at ground level will continue. The situation with sulphur dioxide is less certain.[15] In the long term much will depend upon the pattern of fuel consumption, especially in the domestic sector.

No clear trends can be determined for industrial pollutants such as cement dust, ethylene and fluorides.[10] There is an impression, however, that acute damage has decreased with improvements in industrial pollution control—a trend that should continue.

Vehicle emissions increased by 82–99 per cent (depending on pollutant) between 1960 and 1970.[10,38] Further increases might be anticipated over the next decade in the absence of effective controls unless fuel shortages limit growth in vehicle ownership and fuel consumption. The effects of any increases in emissions are uncertain. Injury to vegetation due to ozone and other pollutants could become more obvious. In the case of lead, much might depend on the contribution of vehicle emissions to total human lead intake.[20] The Department of the Environment is promoting a survey of atmospheric trace metals in urban air in an effort to detect trends in concentrations, especially of lead. The survey will incorporate background data from rural areas.[39]

Emissions of noise from all sources have increased over the last decade.[13,27] Similar trends are anticipated in the future, barring effective controls. Local road traffic noise has been forecast to reach levels sufficient to annoy 35–50 per cent of the urban population by the 1980s.[27] Aircraft noise may increase, although even approximate trends in the distribution and level of noise cannot be determined accurately, given the high rate of technological change in the industry.[12] Noise emissions from all forms of transport could also be greatly affected by the energy supply position over the next decade.

FRESHWATER POLLUTION

Impact
Pollution of fresh water directly affects aquatic biota, water supplies (domestic, industrial, agricultural) and local amenity. There are minor direct and indirect hazards to human health.[40,41] The most common pollutants are bio-degradeable organic matter, suspended solids, ammonia, nutrients, heat and bacteria from industrial, agricultural and sewage waste effluents and land run-off. Many other pollutants such as solids (e.g. wood), detergents, phenols, cyanides, metals, acids, alkalis, pesticides, oils, boron, complex industrial chemicals and biological contaminants (e.g. viruses) may be present. Most of these pollutants are toxic to biota when present in sufficient concentrations.[40,42] Apart from the direct effects of pollution upon fish, indirect damage may occur in some circumstances owing to the obliteration of habitats (e.g. by particulate wastes), encouragement of organisms inimicable to existing biota (e.g. accelerated eutrophication) and destruction of lower organisms serving as food for fish.[40,43]

Acute pollution is confined mostly to waters in urban and industrial areas. According to recent surveys,[44] approximately 25 per cent (by length) of rivers and canals are polluted. This estimate includes many stretches suffering from chronic pollution and classified as doubtful although they still support coarse fisheries. (Some rivers fall naturally into this category and will never support game fisheries.) Unfortunately no national surveys have been conducted to establish conditions in enclosed waters (e.g. lakes, reservoirs).[37,45] Total damage to fisheries is difficult to determine. Commercial and game fisheries have been valued at £2·5–2·8 million per annum,[43] and approximately £125 million per annum is spent by anglers on coarse fishing.[46] Including other forms of recreation, fresh waters must be regarded as assets to society. Thus the elimination of pollution could substantially increase the social value of rivers, canals and lakes.

Nutrients, especially nitrates and phosphates from effluents, run-off and atmospheric fall-out,[47] encourage productivity in oligotrophic or 'nutrient-lean' waters. They accelerate the natural process of enrichment or eutrophication, especially in enclosed waters. Eutrophication encourages the growth of weeds, algae and phytoplankton. Some algal blooms are toxic to fish and mammals. In extreme cases massive growth reduces light availability, and the water becomes deoxygenated when the blooms die and decay. Fish and other biota are excluded from such waters. Although eutrophication has been associated with nutrient levels (e.g. $\geqslant 0\cdot01$ mg P/l water) which are exceeded greatly in many surface waters, the problem seems to be confined in the UK mostly to enclosed waters. This may be associated[43] with many variables, such as high river flow rates and limnological patterns incompatible with severe eutrophication. Nevertheless water undertakings have been estimated to spend £1 million to £2 million per annum to control algal blooms and associated problems in reservoirs.[48]

The hazards of water pollution to public health[49] are recognised by the controlling authorities. The risk of damage appears remarkably low. Rivers and lakes used for public bathing are usually checked regularly for bacterial contamination. Almost all domestic supplies are subject to some routine testing[41,49] to check for pollution at source (e.g. at headwaters) or in system (e.g. contamination of pipelines). Most water undertakings utilise clean waters (e.g. rivers, lakes, boreholes) but some use river water which, because of previous uses, must be subject to cleansing treatments upon abstraction. Pollution (e.g. by pesticides) of reservoirs and boreholes occasionally causes local problems. Where possible the water undertaking will switch to alternative supplies until the pollution is eliminated. However, concern has been expressed about longer-term pollution of groundwaters, which feed boreholes and other sources, by the use of old mines and quarries for the disposal of liquid chemical wastes. (Geologists are studying this problem at the moment.) High nitrate levels in water supplies have been associated with infant methaemoglobinaemia.[48] Nitrate levels do not cause serious problems at present, but concern has been expressed at the rising concentrations in some rivers and groundwaters.[43]

In-system pollution by cross-contamination (e.g. between water and sewage pipelines) is uncommon and transitory. The leaching of metals from metal piping may occur, however, especially in softwater areas. Old lead piping is particularly troublesome, contributing locally substantial additions to the total lead intake of human populations. Replacement of old lead piping with modern materials has greatly reduced the problem. The long-term damage caused by leaching of other metals is not known, despite suggestions that cadmium pollution might be associated with advanced mortality in urban areas.[50]

Trends

Since 1958 there has been a net improvement in the chemical and biological qualities of rivers and canals in this country.[44] This is encouraging in terms of residential and recreational amenities, fishery status and water supplies. Growing demand for water is expected to be the driving force behind future improvements.

However, the pollution criteria used in the national surveys[44] are limited; many pollutants (e.g. pesticides, detergents) are excluded, but apparently exhibit similar downward trends because of improved pollution control. On the other hand, incidents of oil pollution appear to be increasing,[51] with deleterious effects upon local amenities and fresh-water biota. There is also evidence that concentrations of nitrates and phosphates are increasing in some rivers.[43] Generally, however, the obvious damage caused by pollutants such as pesticides, metals and phenols is being restricted more to local incidents of acute pollution associated usually with specific industrial and agricultural operations.

SOIL POLLUTION AND SOLID WASTES

Impact

Soils receive pollutants directly as excess fertilisers and pesticides, accidental spillages of chemicals, etc, and disposals of solid and liquid wastes, and indirectly as inputs from the atmosphere and polluted drainage waters. Some wastes contain pathogens, plant seeds and traces of metals, industrial chemicals and pesticides. Some industrial wastes may be composed almost entirely of toxic or noxious chemicals. Table 2 lists rough estimates[38,52] of some wastes disposed in

TABLE 2 Annual production or disposal of solid and other wastes, 1968–70 [38,52]

Wastes	Approx. weights
Toxic industrial wastes	20 million tons (1–4 million tons extremely unpleasant)
Mine and quarry spoil	65 million tons/annum
Refuse	16–17 million tons/annum
Sewage sludge and animal slurries	140 million tons/annum
Total solid wastes	225 million tons/annum

recent years. No allowance is made for different methods of disposal (e.g. composting, incineration, tipping, sea dumping) which ultimately affect the level of pollution and its impact. Much better information will become available with the establishment of regional Waste Disposal Authorities, who are to prepare detailed disposal plans based upon local surveys.

All wastes have much the same initial effect; when disposed on land they occupy valuable space and obliterate local fauna and flora. Occasionally there are localised problems of pest infestation, wind drift (e.g. odours, dust), contamination of surface waters and temporary sterilisation of soils. Where unpleasant industrial wastes are disposed, toxic substances (e.g. metals) may accumulate in soils in the immediate vicinity of the tipping site, resulting in contamination of plants and, in extreme cases, soil sterilisation and severe water pollution. The main impact of solid wastes disposals, however, is a deterioration of local amenity.

The only indication of the extent of damage is an estimate of derelict land in the UK—approximately 96,700 acres[53]—which includes land severely degraded by all forms of pollution and industrial activity. About two-thirds of derelict land are suitable for restoration by recontouring and revegetation with suitable plant species.[54] The remainder presents more difficult problems of toxicity, waterlogging, etc.

Trends

Trends in solid waste outputs or disposals are difficult to determine and vary greatly between individual wastes.[38] For instance, colliery spoil has increased from 25 to 50 million tons per annum in the period 1955 to 1970,[38,54] despite a fall in coal production, mainly owing to a greater use of lower quality coal seams. China clay and other products of extractive processes have increased in the same period, with certain exceptions such as slate wastes.[38] The weight *per capita* of domestic refuse has not increased greatly over the last decade, but its volume has increased as a result of modern packaging methods.[2] Trends in animal waste, sewage sludge and industrial waste disposal have not been determined. But it is likely that they have increased to varying degrees in recent years. These basic trends in waste *production* are expected to continue. The subsequent disposals are subject to a range of variables of different importance for each waste.[10] For instance, animal waste disposal will be greatly affected by the development of intensive farming techniques and the acceptance of such wastes by waste disposal authorities. Mining and waste tipping are coming under stricter planning controls, often with land restoration as a condition of development. Recycling and by-production are important outlets for some industrial and extractive wastes; both practices are expected to grow in significance as raw materials increase in price. Legislation (e.g. the Disposal of Toxic Wastes Act) and lack of tipping space are encouraging alternatives to tipping and land-fill such as incineration.[38] Thus new forms of pollution (e.g. atmospheric pollution from incinerators) could be generated by the removal of existing problems.

ESTUARINE AND MARINE POLLUTION

Impact

Solid, liquid and atmospheric wastes reach the estuarine and marine environments by direct disposal (e.g. dumping, spillage, discharge), indirect transference in river and drainage waters, and fall-out from the atmosphere.[56] Most pollutants encountered in fresh-water systems can be found in estuaries and marine systems. They have broadly the same direct and indirect (e.g. destruction of habitats) effects upon marine biota, wildlife (including birds) and coastal amenities. However, there are differences in the behaviour and toxicity of pollutants.[42] For instance, salinity affects the solubility of many metals (e.g. mercury is readily transferred from sediments into solution when transported from rivers into marine systems). The environmental stresses faced by receptors are considerably more diverse than those in fresh-water systems and they influence the toxic effects of individual pollutants.

Nevertheless oxygen-demanding substances, suspended solids and nutrients are troublesome in heavily polluted brackish or partially enclosed waters.[56] Local deoxygenation can occur. Industrial disposals (e.g. china clay waste) have obliterated existing habitats in some coastal waters. Algal blooms, sometimes toxic, have been reported off the British coast.[57] Sewage outfalls sometimes contaminate shellfish beds with faecal bacteria, rendering the shellfish unfit for human consumption except after cleansing. Public health regulations exist to protect the public from such hazards.[48,58] Beaches may be polluted with solid wastes (e.g. litter and also offensive sewage solids, with their attendant but often overestimated health hazards). Disposals of sewage sludge and industrial wastes obliterate existing bottom fauna at the disposal site. Occasionally they damage wildlife and pollute local beaches.

A later section deals with persistent pollutants. However, acute pollution by heavy metals, etc, is usually localised around large discharges from specific industries (*cf* Minamata[59]). Persistent pollutants also reach the sea by industrial dumping and as traces in atmospheric fall-out, river outflows, effluents and solid wastes (e.g. sewage sludge). The resulting chronic pollution is reflected in elevated residue levels in wildlife and fish species. The latter are monitored to check that metal concentrations do not exceed acceptable levels in fish for human consumption. Oil, however, is the major pollutant visibly affecting wildlife (e.g. seabirds) and local amenity (e.g. beaches).[50] Unfortunately detergents and other compounds used to treat oil on beaches, etc, are often more toxic to biota than the oil. These pollution problems are essentially local in nature, but they range from chronic pollution in major estuaries (e.g. the Mersey) to episodes of acute pollution (e.g. oil spills). Their impact is greatly influenced by the location of pollution in relation to receptor distribution and the prevailing environmental (e.g. wind, tide) conditions.

The pollution of estuarine and coastal waters has important consequences.[56,61] These areas are used for transport, supply cooling water, serve as major recreational areas, and support shell and offshore fisheries. Coastal and estuarine areas are the breeding, feeding and migrating grounds for many birds and the majority of the world's 'commercial' fish species. Waters around Britain are important to the survival of many commercial deep-sea and coastal fisheries.

The pollution status of our estuaries and coastal waters varies greatly. Some areas (e.g. the Mersey estuary) are severely polluted, whereas others are surprisingly clean, considering the demands of local populations and industry for disposal and resource (e.g. amenity) facilities.[62] Evidence of damage to fisheries of all kinds is hard to ascertain, given their sensitivity to natural stresses (e.g. storms) and over-fishing. Estuarine and marine ecosystems exhibit great natural changes over time in the absence of pollution. Nevertheless considerable numbers of seabirds are killed every year, mainly by oil pollution[50] but occasionally by ingestion of, or entanglement with, plastics and other pollutants. Similarly there is extensive fouling of beaches with oil, solid wastes (including sewage, coal spoil, etc) although this is highly variable in location and in frequency.

Trends

Reports[44,62,63] reveal that in tidal waters, including estuaries, conditions generally deteriorated between 1958 and 1970, although there has been a slight improvement recently. This trend hides extremes in which some waters have remained grossly polluted whilst others have shown spectacular improvements in quality (e.g. the Thames estuary). Generally, data for coastal waters are sparse and give only the barest indication of trends. For instance, there appears to have been no general increase in metal concentrations.[64] There is circumstantial evidence that oil pollution is increasing, and disposals of sewage, industrial wastes and cooling waters are also rising. In future much will depend upon the efficiency of voluntary and legislative control measures, reinforced by improved monitoring systems, as well as improvements in technical controls (e.g. the prevention of oil tank washings being discharged at sea). The location of industry and its demands for sea disposal and other facilities will remain the dominant factor influencing future trends in pollution.

PERSISTENT POLLUTANTS

These include plastics, metals, fluorides, radioactive wastes and certain pesticides and industrial chemicals. They are discussed separately because of their ability to remain in the environment, often with little or no degradation, and to be transported through more than one medium by physico-chemical and biological diffusion processes.

Impact

Plastics　Plastics are usually considered as solid wastes (e.g. litter, refuse) which are not amenable to the usual degradation processes. They degrade local amenity (e.g. litter) and occasionally cause the death of farm animals and wildlife, and local nuisances to industry (e.g. fouling of fishing nets[57]). The most serious problem, however, is that of effective disposal. Land-fill and marine dumping are often unsatisfactory because plastic wastes are so bulky and non-degradable. Incineration, if carried out properly may be an attractive alternative, and some types of plastics can be recycled if segregated from contaminants before treatment.

Pesticides　In 1966–67 approximately 600 tons of organochlorine pesticides were used in this country, primarily for crop protection, although 10 per cent were used in industry (e.g. mothproofing).[65] There are several examples of acute pollution by organochlorines, with lethal results for wildlife, especially birds (e.g. Clear Lake[66]). The death of many birds in this country in the 1950s and 1960s due to the use of Dieldrin as a seed dressing resulted in the establishment of the Pesticides Safety Precautions Scheme between industry and government, which is administered by the Advisory Committee on Pesticides and Other Toxic Chemicals used in Agriculture. Currently there is little sound evidence that man or any other animals are suffering significant damage from this form of chronic pollution, with the possible exceptions of some birds.

Research has demonstrated that very low levels of environmental contamination may have serious consequences where organisms concentrate pesticide residues in their tissues.[67] Further concentration may occur as the residues pass along food chains. Thus, the highest trophic orders (e.g. porpoises, carnivorous birds, man) can accumulate residues at places remote from the original site of contamination in space and time, and occasionally in sufficient quantities to cause themselves damage. Some predatory and fish-eating species of birds have suffered considerable declines in population.[67,68] These declines have been associated with high levels of pesticide residues, which are believed to affect breeding success by causing reductions in eggshell thickness and other physiological disorders. Certainly the current low status of some British bird populations may be associated with past patterns of pesticide use in agriculture.

PCBs Polychlorinated biphenyls (PCBs) are less acutely toxic than organochlorine pesticides. But they are concentrated in organisms and have been associated with damage to wildlife, especially to birds.[68, 69] PCBs have been detected in sewage sludges, surface waters and animal tissues. It is suggested[70] that some PCB pollution is derived from fall-out of photo-oxidation products of atmospheric organochlorines (e.g. DDT). However, most PCBs probably originate in effluents, etc, and obsolete products from secondary manufacturers and consumers, especially of electrical machinery.[71] The PCB problem illustrates the potential hazards of using persistent chemicals in industry without adequate testing procedures for toxicity and environmental behaviour.

Metals Many trace metals are known to accumulate in waters, sediments and soils and to be concentrated by plants and animals. They also may be concentrated as 'residues' in receptors as they pass along food chains. Severe pollution and acute damage are rare, being localised around poorly controlled disposal sites. Classic examples include the Minamata (mercury)[59] and Itai (cadmium) incidents in Japan. Under normal circumstances, however, there is no evidence of overt damage to man[72, 73] or to animals by transmission of metals along food chains. Nevertheless high body burdens of mercury have been implicated in breeding failures, etc, among some bird populations. Generally the situation is complicated by the fact that animals, especially man, receive metals by various routes from the environment (e.g. lead in food, water, utensils and air). Also the effects of individual metals must be unravelled from a complex web of interactions with other trace metals, the intake of which depends greatly upon the diet of the receptor.

Radioactivity Radioactive pollution is more intensively monitored than any other form of pollution. Routine or spot checks are conducted in all media, especially near industrial sites utilising and disposing of radioactive materials, and in samples of food, drink, water and human 'tissues' (e.g. bone analyses).[12] Background contamination is low and dominated by medical and natural exposures except during periods following nuclear weapons tests, when these sources may be swamped by radioactive fall-out. Occupational and waste-disposal exposures are usually negligible. The total addition by man to background radiation is small and cannot be associated readily with any somatic, genetic or other forms of damage. However, the very long-term significance of man-made radiation to human health is the subject of some dispute.[74]

Monitoring of major routine discharges of radioactive wastes is sophisticated and involves the determination of the critical pathways of hazardous pollutants through the environment to the most exposed receptor, with the emphasis upon man (chapter 6).[75] This procedure requires the identification of food chains and concentration factors at various trophic levels. The advantage of such an approach is that the maximum permissible discharge can be calculated from the maximum exposures considered safe to the receptor. The best example may be found in the studies of the Windscale discharge to the Irish Sea.[75, 76] No damage to man or to wildlife, etc, can be attributed to existing radioactive discharges operating under normal conditions. Exceptions could arise, however, if accidents causing short-term acute pollution were to occur.

Since the Windscale incident, in which the surrounding countryside was contaminated by an accidental, short-term radioactive emission, considerable attention has been paid to the theoretical risks of pollution due to accidents in nuclear power stations (chapter 6).[77] This concern has recently increased in the UK, owing to government consideration of alternative future power reactor systems. Studies of the failure characteristics of reactors and their components place the probability of a major accident at 10^4 reactor years for a release of radioactivity equivalent to 10^4 curies 131-iodine (the estimated primary damage is thirty-three cases of thyroid cancer in a typical reactor location). For comparison, world reactor experience is estimated at 10^3 reactor years by the year 2000.[78]

Apart from the gaseous and liquid low-level wastes, some radioactive solids (e.g. glassware) are disposed of at sea or on land.[12, 79] These are small in quantity and are not detectable as increases in environmental radioactivity levels. Most radioactive wastes (99·9 per cent total activity)—'high-level' wastes—come from nuclear power fuel processing. These are stored in solution in secure, cooled tanks. Few such wastes have leaked to the environment, although the required period of storage to reduce radioactivity to levels currently considered environmentally acceptable is extremely large (about 1,000 years' duration).

Fluorine   Around certain industries (e.g. brickworks) particulate and gaseous fluorides are emitted to the atmosphere and deposited upon vegetation in the immediately surrounding countryside. If eaten by farm stock and wild animals this contaminated vegetation can cause fluorosis; fluorine is accumulated in the teeth and bones of animals, resulting in loss of hair, appetite and milk yield, dental lesions and eventually paralysis or crippling fluorosis.[32] Doses may be lethal in rare cases; these, however, are strictly local phenomena.

Fluoridation of water has caused concern in some quarters, mainly because of the narrow margin between the doses necessary for prevention of dental caries in children and those believed sufficient over very many years to cause fluorosis in adults.[80] There is no reliable evidence of damage, and the long-term effects of fluoridation are open to speculation.

Trends
Plastics   The major growth of plastics has been in packaging, which is reflected in the rising trends in bulk and plastics content of litter and refuse over the last decade.[10,55]. Although these trends may well continue in the future, the cost of raw materials, changes in packaging techniques (e.g. degradable plastics, returnable containers) and disposal methods (e.g. substitution of incineration for tipping, and greater recycling of plastics) could greatly reduce these increases. However, there are technical problems to be overcome in recycling (e.g. reformation of complex plastics) and incineration (e.g. corrosion of incinerators, air pollution by noxious gases).

Pesticides   In the 1950–60 period organochlorines (e.g. DDT, Dieldrin) were used extensively in agriculture for cereal seed dressings.[65] This resulted not only in cases of acute toxicity to many birds but also in declines in the populations of many raptorial species.[68] The Pesticides Safety Precautions Scheme, a voluntary system for controlling pesticide pollution in agriculture, brought this problem under better control by restricting the types of pesticides in use and their methods of application. It also tended to encourage the substitution of less persistent pesticides (e.g. organophosphates). Currently, serious damage to wildlife is minimal, with some exceptions, probably due to misuse of pesticides.[81] Within the last five years data on residues in foodstuffs, human diet, human fat and various species of wildlife have all exhibited downward trends of varying magnitude.[12,32,72] There is circumstantial evidence to suggest that organochlorine pesticide pollution of rivers, due mainly to industrial and agricultural effluents, has decreased, although better control is advocated in this area.[83] Innovations (e.g. persistent herbicides, slow-release pellets of pesticides) may create new ecological problems, especially in terms of water pollution.

PCBs   There are insufficient data on PCBs to determine trends in pollution. However, their persistency in the environment and probable association with breeding failures among birds and possibly other wildlife[68, 69] have resulted in national and international moves to withdraw PCBs from use except for essential purposes and in closed-circuit systems.[71]

Radioactivity   Data on radioactive waste disposal and pollution are available only in certain cases and require careful interpretation. General environmental levels of radioactivity reflect primarily fall-out from atomic weapon tests. Thus levels of a variety of radioactive elements in surface coastal waters, rivers, rainfall, drinking water, foods (including milk) and human bone were nearly all at their peak in the period 1962–64, having since declined to about their pre-1960 levels.[12]

Data are not available for local contamination by atmospheric emissions from establishments using radioactive materials. Disposals of solid wastes, however, are recorded. Inland burials were variable over the period 1964–70, as were burials at sea, which ranged from 1,700 to 75,000 β activity irrespective of tonnage dumped.[75, 79] Discharges to inland waters varied with the works involved, but no significant increases occurred over the same period of time. Discharges to coastal waters are dominated by the Windscale effluent, which increased from 15,200 to 27,430 β activity and from 71 to 767 α activity in the period 1964–70.[75] The doses received by receptors exposed to the resulting pollution[75] do not exceed current international standards for radiological protection. No obvious pollution damage can be detected.

Undoubtedly the major sources of radioactive wastes now and in the future are nuclear power generation and its attendant fuel industry. Nuclear power could account for 60–80 per cent of UK energy consumption by the year 2000.[84] This would result in the emission of greater quantities of low-level wastes and the generation of greater quantities of high-level wastes than at present, although the actual quantities will be determined by the type and efficiency of the reactors in operation. Of the low-level wastes, 85-krypton might reach sufficiently high concentrations in the atmosphere by the year 2000 to exceed recommended safety limits for exposure to radiation in certain industries (e.g. those extracting krypton from the atmosphere) and possibly some local populations. Similar problems could arise with tritium (as tritiated water) and other wastes at still later dates. The remedy may lie in the temporary storage of such wastes to reduce activity before they are discharged to the environment.[84]

High-level radioactive wastes in solution currently require considerable investment in space and equipment for long-term storage (up to 1,000 years). To reduce the storage costs and the hazards inherent in accidental releases of wastes to the environment most nations are considering solidification of high-level wastes.[85] For instance, insoluble refractive blocks would hold many curies of waste energy, requiring only air-cooled and secure housing instead of sophisticated tank complexes. Some increase in the probability of a serious polluting accident to a nuclear power reactor is considered inevitable with any increase in the size of the industry both here and abroad.[78]

The extent of damage and its actual occurrence will depend not only on the growth of the industry but also on the type and efficiency of the reactors, their safety technologies, and their location in relation to centres of population. With a massive growth rate of the industry based on current technology, the theoretical probability of a major accident in the year 2000 causing eventually some thirty deaths due to thyroid cancer is lower than almost any other man-made (e.g. plane travel) or natural (e.g. lethal strike by a meteor) hazard. Possibly the development of the breeder and fusion reactors will eliminate many pollution problems, but it is equally likely that they will present new environmental hazards.

Metals   There are few time-series data available to determine trends in metal pollution—a situation which should be remedied soon.[14] There is no sound evidence that general urban lead pollution has increased significantly over the last decade, despite increases in emissions. (Local exceptions have been observed close to busy roadways.) This may be due in some way to a balance between increasing vehicular emissions of lead and decreasing contributions from industrial, solid fuel and other sources. Surveys of British coastal waters for several metals indicate no significant trends in recent years except in certain estuaries where increases (e.g. in cadmium) can be associated with specific industrial activities.[64]

Fluorine   There is circumstantial evidence that local pollution by fluorides is declining with technological changes and better pollution control in various industries (e.g. brickmaking, iron and steel making, pottery production), although there are some new sources (e.g. aluminium smelting).[86]

GENERAL COMMENTS

Our knowledge of pollution is deficient in most areas, few more so than in recent trends in pollution and pollution damage. We have little idea of the damage incurred by populations and communities of receptors. The effects of continual exposure to low levels of such pollutants are subtle but potentially of great long-term significance. It is significant that research workers[87] are now considering pollution as only one environmental stress upon receptors rather than as an independent causal agent in disease, disorder and death.

Despite lack of data, it is clear that the *general* view of pollution as a growing menace is untenable. In most cases acute forms of pollution seem to be declining. This is also true of many forms of chronic pollution, although examples of increasing and fluctuating pollutant trends are known or suspected. We need improved monitoring systems for all types of pollution. In future damage is expected to devolve into (*a*) widespread, sub-lethal responses to chronic, very low-level pollution; and (*b*) very localised, severe responses to limited, acute pollution arising from accidents (e.g. to oil tankers) and poor waste disposal techniques.

CHAPTER 3

Global pollution: a summary

It is difficult to grasp the global significance of a particular pollutant partly owing to the scale of the problem and partly owing to the uncertain state of knowledge about global pollution itself, a consideration which has contributed greatly to recent demands for global monitoring systems.[88, 89] An attempt is made in this chapter to discuss the major forms of global pollution and their probable consequences.

There are four basic characteristics of a global pollutant:

1   It is discharged into media (e.g. atmosphere, oceans) within and/or between which dispersion and mixing are widespread and relatively rapid.
2   It is released at many points throughout the world.
3   Its effects are widespread within one or more media.
4   It is often persistent and accumulated in the environment and in biota.

In principle, a complete global pollutant is discharged to the environment at many points around the world, and is accumulated in the environment and in biota with widespread actual or potential adverse effects.

In practise, global pollutants rarely exhibit all four characteristics, although most are discharged at many points around the world into media within and between which mixing and dispersion occur. They fall, instead, into two main groups. The first consists almost entirely of pollutants readily dispersed in the atmosphere, with adverse effects upon global climate and, indirectly, upon global ecosystems. The second group consists of pollutants with direct ecological effects. They may be persistent and accumulated within biota. In most cases their global impact stems from a widespread incidence of essentially local ecological effects in the marine environment. There are exceptions to this pattern. For instance, some pollutants (e.g. DDT, lead) are widely dispersed throughout several media, including the terrestrial environment, but cause only local, overt damage where concentrations are exceptionally high.

It is not possible to cover all aspects of individual pollutants. In common with other reports,[88] this chapter considers briefly the principal aspects of pollution by carbon dioxide, sulphur dioxide, nitrogen oxides and other trace gases, particulates and aerosols, heat, jet transport, radioactive materials, alterations in land use, insecticides, polychlorinated biphenyls, metals, nutrients, and oils and associated materials.

CLIMATIC EFFECTS

The results of changes in the earth's surface (e.g. construction of dams and cities, removal of forests) are not regarded here as pollution *per se,* although they have direct effects upon climate.[88]

Thus, the change from forest to crops increases the albedo of the earth's surface, decreases the availability of water to the atmosphere, and decreases surface roughness, with consequential effects upon turbulence, evaporation and wind speed. Large urban areas create heat islands which produce approximately 10 per cent of the equivalent total solar heat absorbed at ground level and cause shifts in the speed, updraft and turbulence of winds. Total urban thermal output has been forecast to increase by up to 600 per cent of the present level,[91] but for each urban unit and for the world as a whole this will not significantly elevate general temperatures.

Carbon dioxide

Carbon dioxide is a natural constituent of the atmosphere, at about 320 ppm, having increased from 310 ppm at an average of 0·2 per cent pa since the late 1950s. This increase may be correlated with increased combustion of fossil fuels—a relationship corroborated by slightly higher average levels (especially during and shortly after winter months) in the northern hemisphere, where fuel consumption is greatest. Further support[88] may be found in an observed 1·2 per cent decrease between 1850 and 1950 in the radioactive $14-C$ content (from natural sources) of biological specimens, which is attributed to a proportionate increase in outputs of CO_2 from fuel (lower $14-C$ content) combustion in the same period.

The possible effects of this additional CO_2 depend on the capacity of the environment to hold CO_2 and to convert it to other substances (e.g. by photosynthesis). In the last decade about 50 per cent of the CO_2 produced has remained in the atmosphere, the rest being distributed between the biosphere (all organic matter) and the oceans.[88]

The total capacity of the former is difficult to assess, but it is believed to contain one to two times the total atmospheric carbon. In theory this could lead to a corresponding increase in the growth of plants, but the effect is not observable. The oceans contain about sixty times as much CO_2 as the atmosphere. However, only the first few hundred metres are important because of the slow turnover rate of deeper waters. About 1 per cent oceanic CO_2 is in solution, the rest being in the form of carbonates and bicarbonates. Between them the biosphere and oceans may contain total CO_2 reservoirs three to nine times greater than the content of the atmosphere. The *accessible* oceanic reservoir may be only a fraction (30–80 per cent) of the latter.

Forecasts of global fuel consumption indicate that man *could* increase the CO_2 of the atmosphere by 20 per cent by the year 2000[88] and by a factor of four or more in the next century, assuming that current patterns of consumption are maintained. CO_2 is important in the determination of the temperature of the planet by absorbing a limited amount of sunlight and by absorbing infra-red radiation. The CO_2 component of the atmosphere also emits some infra-red radiation. Any increase in CO_2 could, therefore, result in an increase in the mean surface temperature—the 'greenhouse effect'.

Extreme increases in global temperatures would be catastrophic, melting the ice caps and causing widespread flooding of lowland areas, where many of the world's urban centres lie, disrupting agriculture, industry and domestic life, and changing whole ecological communities beyond recognition. The extent of the damage, including loss of human life, would depend greatly on the speed of heating reactions.

One calculation indicates[88] that an 18 per cent increase in CO_2 by the year 2000 would increase the surface temperature 0·5 per cent and cool the stratosphere by 0·5–1·0 per cent. It must be remembered, however, that many estimates of temperature changes make no allowance for interactions between atmospheric dusts, aerosols, etc, and the complex thermodynamics of the atmosphere. Others omit important factors such as vapour absorption, condensation, surface evaporation and convection, cloud absorption, etc. Furthermore there is no evidence of any significant recent increase in the earth's surface temperature. Until all radiative and dynamic features can be combined into a single forecasting model the 'greenhouse effect' remains a speculative but sombre reminder of one possible remote consequence of human activities.

Sulphur dioxide

A substantial proportion of SO_2 is derived from natural sources whose total output may vary substantially but temporarily with specific natural phenomena (e.g. volcanic eruptions).[15,88] Although SO_2 causes considerable local problems it is not generally regarded as a serious global toxicant. However, there is much discussion about the possible effects of SO_2 emitted in Britain upon the forests and lakes of Scandinavia.[92] Work is proceeding in this area, but it is often not realised that the origin of the acidic rainfall (acid/sulphate effect—not direct gas effect) is not entirely consistent with the assumption of prevailing south-westerly winds from Britain, because many of the winds over Scandinavia come from heavily polluted areas of Germany and central Europe.

The removal of SO_2 from the atmosphere[15,88] is complex, including the processes of photo-oxidation, catalytic conversion, chemical combination, impaction, sedimentation, wash-out and rain-out. The average turnover time for SO_2 in the atmosphere is about forty-two days, with a mean molecular lifetime of one to four days—features not conducive to environmental accumulation. However, total emissions both here and abroad are increasing, especially as high-level emissions (e.g. from power stations); SO_2 is dispersed and converted to other substances at much higher altitudes and far greater distances before returning to ground level. The immediate effects of SO_2 as a toxicant are thus alleviated. The formation of particles and aerosols from discharged SO_2 with other materials, however, has global implications, the most important of which is believed to be its contribution to the global atmospheric loading of aerosols and particles which influence radiation. This applies to all other gases which man adds in varying proportions to the natural loadings already present in the atmosphere. For instance, natural hydrogen sulphide contributes approximately 30 per cent more sulphur to the atmosphere than man-made sources.[15, 88] Similarly, nitrogen oxides from polluting sources contribute less than 2 per cent of the total nitrogen, which is mostly derived from natural sources.

Particulates (including aerosols)

The sources of tropospheric particulates[88] include:

1 Natural aerosols from dust storms, etc (approximately 0·3 μ particle radius); photochemical reactions between ozone and natural hydrocarbons (<0·2 μ radius); photochemical reactions between

   various natural trace gases such as SO_2 and ozone; volcanic eruptions emitting particles and gases; and possibly by evaporation
and condensation from plants and soils.

2   Natural aerosols from sea sprays and other water sprays.

3   Man-made particulates from industry, domestic smoke and
photochemical reactions involving man-made trace gases (e.g.
SO_2, hydrocarbons) which augment natural reactions.

Most water vapour for the formation of aerosols from gases and dry
particulates comes from natural sources. It is also possible that condensation from industrial cooling units could have a locally significant
effect upon atmospheric vapour levels in the future. The evidence for
man-made alterations of the natural balance of stratospheric contamination lies in the discovery of stable lead particles, and the effects
of nuclear tests and volcanic eruptions on the particulate populations
in the stratosphere. The most important activity of such particles is
their role as condensation nuclei for water vapour.[15, 88] This is the
mechanism by which most clouds are formed; precipitation (rainfall)
depends largely on the number of nuclei available in the original cloud-
forming updraft. It is also possible that particulates serve as nuclei
which initiate the freezing-out of supercooled water droplets. Both
processes increase cloudiness and also possibly increase precipitation.

   Cloud and atmospheric particles absorb, attenuate and scatter solar
radiation.[88] Scattering is predominantly downward, but the 10 per cent
upward scattering tends to increase the reflectivity of the particulate
layer. Back-scattering (reduces heating) and absorption (increases
heating) tend to counteract each other below this layer and at ground
level. A warming effect, however, may be obtained with relatively
small absorption and back-scattering in relation to high magnitude re-
flection from the earth's surface. At high altitudes cooling will occur
irrespective of the warming effect. Prevailing trends of increased tur-
bidity (particulate density), especially in the northern hemisphere,
have been associated with a decrease in solar radiation over the last
thirty or forty years. The overall effect of lowering surface temperatures could ultimately lead to a second ice age (at the very worst)
running through the gamut of more temperate conditions, lower sea
levels, and increased cloud cover and precipation, all with attendant
effects upon man and other organisms. The total effect could be
one of cooling of the surface in direct contrast to the 'greenhouse effect'
caused by CO_2 accumulation. These two effects may well be masking

in the next century will be great; they will probably fall within the range of normal global temperature variation.

Jet transportation

This contributes pollution to the atmosphere[88] by (*a*) the formation of condensation trails by subsonic aircraft (mostly in the upper troposphere) and (*b*) the contamination of the stratosphere by supersonic jet emissions especially from transports (SSTs) such as the 'Concorde'. There are tentative indications that cirrus cloudiness has increased recently in certain areas with the increase in jet flights. This would contribute to the cloud/climate effect, (e.g. decreasing solar radiation at ground level and increasing the reflectivity of the cloud layer).

SST exhausts could interfere with the stratosphere in two main ways. Firstly, aerosol emission components (e.g. SO_2, NO_x, hydrocarbons, soot), carbon dioxide and water vapour are retained in the stratosphere for some time. Theoretically they could interfere with the stratosphere's heat balance, but such effects are considered minute compared with those mentioned already for CO_2 and particulates. Secondly, photochemical oxidation of stratosphere ozone by nitrogen oxides and water vapour from SST exhausts could reduce the earth's ozone shield against harmful ultra-violet radiation. However, the stratosphere contains both contaminants in far greater quantities than could be introduced by SST flights. The overall effect of SST operations is expected to fall within the range of natural global variation.

Heat

Heat released by combustive processes may be regarded as climatic contamination. Heat islands over cities are a manifestation of this effect.[88,91] Heat is released in two forms: sensible and latent. The latter is associated mainly with the dissipation of heat in air, surface waters and *via* cooling towers during power generation. Latent heat lost by evaporation to the atmosphere is estimated at 15–20 per cent of total waste heat, but it is considered available to the atmosphere at much higher altitudes than sensible heat. The overall heat loss and vapour emissions are not thought to be significant in global terms. However, future increases in power generation requirements could create severe local demands on water resources for cooling purposes, particularly in tropical and semi-tropical areas, leading to considerable water loss by evaporation and to thermal alterations of local aquatic ecosystems. Any contributions to cloudiness are uncertain but will probably be insignificant except locally.

ECOLOGICAL EFFECTS

The ecological effects of pollution, other than those due to climatic change, are not all considered in detail here. Most of the global problems on land and in fresh water either arise from man's direct demands upon resources or are essentially local pollution problems (these may, however, be common throughout the world). The major potential global ecological effects arising from projected alterations of the world's surface and changes in methods of resource utilisation can be summarised as:

1 Selective impairment of organisms by pollutants (e.g. heat, metals and pesticides). Among populations of insects and other arthropods this can lead to outbreads of pests in agricultural areas. Damage may also be caused by the inhibition or destruction of lower sections of food webs (e.g. plants and herbivores).

2 Selective impairment of organisms at specific stages in their development (e.g. the higher sensitivity to pollutants of the egg and larval or juvenile stages than that of adults). This may lead to the failure of a species to sustain a stable or expanding population.

3 Reductions in the stability of ecosystems in areas where natural communities are replaced by mono-crop cultures. Mono-crop cultures contain fewer species than most natural communities, creating instability of the ecosystem. The use of pesticides intensifies such phenomena.

4 Reductions in the number and range of species are associated with the attrition of ecosystems by human development, and with pollution of the terrestrial, fresh-water and marine environments.

5 Damage to insects, etc, involved in natural pollinatory systems of vegetation. These can occur in areas of high atmsopheric pollution and excessive pesticide application.

6 Damage to inshore breeding grounds for much of the world's fish fauna[56, 61]—already a problem in respect of river/estuarine pollution, over-fishing, inshore waste disposals, metals, pesticides, polychlorinated biphenyls, oils and spoil disposals.

7 Potential long-term genetic damage to future generations of humans and other species. Radioactive wastes are prominent in arguments about the probability of such damage, but any chemical mutagen, carcinogen, etc, should be regarded with caution, especially if persistent and accumulative.

8 Climatic effects of land surface alterations and general human

activity, including removal of forest cover, urban development, dust and aerosols created by industrial and other activities, etc, all of which have attendant ecological consequences.

9   Impairment of soil condition. Severe local problems occur owing to compaction by heavy machinery and erosion by rain, flood and winds. Widespread impairment of soil micro-organisms by persistent pollutants would not only impair the soil creation process but also the basic cycling of nutrients within ecosystems.

Some of these changes are detectable already (e.g. reductions in the diversity of ecosystems) but others appear much more remote. Potentially, the most significant global changes are associated with pollution of the marine environment. The following sections discuss the sources, environmental pathways and effects of persistent radioactive materials, pesticides, polychlorinated biphenyls and metals. There are also brief sections dealing with nutrients and oils.

Radiation and radioactive materials

The global effects of wastes from conventional nuclear power production and fuel processing appear to be non-existent[74,84] for estimates of production up to the year 2000. This ignores fall out from atomic weapons and accidents to reactors. The risks of reactor accidents are remarkably small, but the increasing use of reactors inevitably increases the chances of a serious accident.[77]

Nuclear power production generates over 200 radioactive substances, including radioactive isotopes such as 131-iodine and the noble, inert gases (e.g. argon, xenon and krypton). Two products, krypton (gas) and tritium (as tritiated water), are released into the environment in relatively large quantities. They have relatively long half-lives (greater than ten years) and therefore will increase in concentration in the environment if the discharge rate exceeds the rate of decay. Tritium (because of its dilution in water and difficult separation from the medium) and krypton (because of its inertness) are not easily removed from nuclear waste streams. Technical solutions to the problems of extracting and storing these wastes are under active investigation.[81] Nevertheless it has been estimated that by the beginning of the next century 85-krypton will be sufficiently concentrated in the atmosphere to produce doses at or above the recommended safety levels[84] where the gas is extracted and concentrated for industrial purposes. At still later dates tritium and noble gases such as xenon and argon could create similar problems.

The environmental importance of nuclear wastes in the future rests on: (*a*) nuclear power production increasing in this country and abroad, and (*b*) waste production and accident hazards inherent in new types of reactors. However, reactors normally provide only a small fraction of the total exposure to radiation compared with doses from natural (cosmic and background), medical (x-rays, therapy), occupational and military sources.[29] All except natural sources are as closely controlled as is currently believed to be practicable in terms of public safety and control costs. For instance, control procedures for wastes discharged to the sea include the monitoring of concentrations of radioactive materials in organisms, their transmission through food chains, and the doses ultimately received by man.[75] Nuclear power plants in this country produce less waste than their maximum authorised discharges, which are set below the recommended limits for exposed human populations.[75] Nevertheless some scientists regard any additional radiation as dangerous and all existing standards as far too weak.[77] The evidence for such arguments is inconclusive, but the possibility of subtle, low-level somatic and genetic effects cannot be entirely dismissed.

The problems of storing 'high-level' liquid nuclear wastes (99·9 per cent of all nuclear wastes) are causing concern, primarily because of the quantities involved and the difficulties of long-term containment. A major accident at a storage site anywhere in the world might have serious implications for aquatic and marine life—even on a global scale. To reduce such risks, and to minimise storage space and maintenance overheads, most countries are considering the solidification of liquid wastes (e.g. as insoluble, refractive blocks) within the next decade.[85] In the longer term the fast breeder reactor should lead to the more efficient use of fuels and therefore less waste. It is feasible that the practical application of fusion reactor theory could eliminate the production of many wastes. These changes are not necessarily devoid of their own environmental problems.

Pesticides
The most widespread, persistent organic insecticides are the organochlorines, especially DDT and its derivatives (e.g. DDE). DDT is used to protect man from disease (e.g. anti-mosquito sprays) and deprivation (e.g. protection of crops, stored products and some materials) throughout the world, with enormous material and economic benefits. Other organochlorines, such as Dieldrin, have

been used quite extensively. The major source of contamination is agriculture, although domestic, medical and industrial users make smaller but substantial contributions to the total environmental load.[65,88]

The majority of organochlorines used eventually arrive in the marine environment. The prevailing concentration of DDT in the sea is generally low (i.e. a few parts per trillion).[88] But marine animals tend to concentrate organochlorine; even plankton at the base of food webs contain about 0·01 per cent ppm DDT residues. Higher up food chains fish average less than 1·0 ppm, although higher peak values are not uncommon (e.g. 6 ppm in whales, 18 ppm in porpoise, 11 ppm in gonads of tuna, 5·4 ppm in oysters, and 10 ppm in the fat of the many marine animals, including seabirds).[88,93] There is great variation in observed concentrations between sample sites, individuals from the same sites, tissues in the same individuals, and in techniques of sampling and measurement.[94] Apart from high local and individual values, however, low levels of residues (0·001–1·0 ppm) can be found in marine and aquatic animals throughout the world. Even Antarctica is contaminated. Residues have been observed in penguins and in snow (about 40 ppt). These appear to establish the global distribution of the pollutant.

The total capacity of the oceans is estimated at $7·5 \times 10^7$ metric tons of DDT in solution, equivalent to approximately ten times the estimated world production pre-1970.[88] Within the oceans DDT has a residence time of several years, but there is no evidence of uniform distribution even in the mixed layer (0–100 m depth). Local discharges of DDT into ground and river waters and in solid disposals (containers, sludges) are highly variable. The annual global surface run-off of water from places of application on land and in inland waters has been estimated to contain $3·8 \times 10^3$ metric tons of DDT (equivalent to about 1×10^{-3} of annual world production).[88,95] The overall contribution may be higher, since most estimates omit residues bonded on to sediments moving to the sea through river and drainage systems.

The airborne input of organochlorines in aerosols is uncertain. The residence time of DDT in the atmosphere is probably only a few days. Direct adsorption into water may be assisted by the occurrence of thin surface films of oils and fatty substances.[88] Rainwater has been measured at 80 ppt DDT in the UK and up to 1,000 ppt elsewhere.[65] If world precipitation averaged 80 ppt the total deposition into the oceans would be equivalent to about 25 per cent of global annual production. Other sources such as sludge disposals contribute an unknown but probably significant amount of DDT to the overall marine loading.[95] The total quantity of DDT in the marine environment is, therefore, roughly estimated to be equivalent to 40–60 per cent annual production. A limited quantity of the remainder resides in the atmosphere, and the rest in inland waters, soils and terrestrial organisms.[88]

Only 0·1 per cent of total production is reckoned to exist in marine biota. However, residual levels similar to those reported earlier have been associated with damage to some animals.[88] Low-level pollution may affect the survival of some plankton species. Reproductive inhibition or failure in some species of fish and crustacea has been reported at 10 ppt–0·1 ppb DDT in seawater. Fish residues of 10 ppm have been associated with the disappearance of fish-eating birds. Evidence of various biochemical effects of DDT contamination is accumulating. Lack of breeding success and death amongst avian populations contaminated with organochlorines are well documented, although the association between organochlorine contamination and eggshell thinning has been disputed.[68, 96]

Acute pollution by pesticides is generally regarded as declining in incidence. Low-level chronic pollution, which has sub-lethal effects on animal breeding and behaviour, gives some cause for alarm because of its likely widespread occurrence. Many coastal and estuarine areas are already too polluted for comfort. With an estimated turnover of about five years in the ocean and an unknown percentage of the remaining 75 per cent of total DDT production to date yet to reach the oceans, it is clear that the situation cannot be alleviated rapidly even if the production and use of DDT and other organochlorines ceased immediately.[88] This is impracticable because developing countries in particular cannot afford to forego the benefits of DDT until less dangerous products of equal cost effectiveness are provided. Nevertheless there are some substitutes available for certain purposes. Greater care could be exercised to avoid over-use and misapplication. In various countries official and voluntary restrictions are coming into force with such objectives in mind. In this country residual levels in foods (including fish), wildlife and humans are showing generally encouraging downward trends[72, 81] in response to pesticide control measures (chapter 2).

Carbamate (e.g. Carbamyl) and organophosphate pesticides are replacing organochlorines for many purposes. Their routes of entry to the seas may be similar to those of organochlorines. But although many are highly toxic and can cause severe local damage, they are much less stable, far less persistent and non-accumulative. They are therefore unlikely to be as widely dispersed in the environment as DDT.

Polychlorinated biphenyl compounds (PCBs)

PCBs have many characteristics in common with organochlorine pesticides; they are often found on the same animal tissues during pesticide residues analyses.[72] They are produced in many formulations and have a wide range of uses, especially in the paint and electrical industries. They are present in most municipal sewage sludges and probably in precipitation from the atmosphere. It is likely, however, that most PCBs are released by vaporisation and in various effluents during their incorporation into manufactured products. The use of such products, and their eventual disposal when obsolete, are likely secondary sources of environmental contamination. High levels of PCB residues in birds have been recorded in this country.[69] Reports from elsewhere (e.g. Sweden and the USA) indicate incipient if not actual global dispersal of PCBs similar to that of DDT. One report suggests that mid-ocean concentrations are higher than those in coastal waters,[97] which may be due to the fall-out of PCBs formed by photo-oxidation of organochlorines (e.g. DDT) in the atmosphere over the oceans.[70]

PCBs are less acutely toxic than organochlorines. There is no evidence of acute toxicity to wildlife in the field similar to past incidents with Dieldrin, etc. However, PCBs are concentrated by some organisms and passed along food chains. They have been implicated in bird deaths[87] and are believed to impair avian breeding success;[68] PCBs may be important factors in the observed decline of some predatory species of birds. Subtle damage to populations of fish, crustacea and other marine organisms cannot be ruled out.

Trends in environmental levels of PCBs are difficult to determine. Nevertheless their widespread distribution and the subsequent indications of hazards to wildlife have been sufficient to induce attempts at close control of their sale and use. The withdrawal of PCBs from the market and the substitution of less dangerous materials may be the only completely effective solution.[71]

Comparable problems could well arise with some of the many new industrial chemicals produced each year. None of these is subject to routine tests for toxicity and persistency similar to those applied to agricultural, medical and veterinary chemicals.

Metals

Several trace metals can be regarded as potential global pollutants. In particular mercury, lead and cadmium deserve mention because of the widespread publicity (*cf.* Minamata) given to their uses and effects.

Mercury contamination of inland and coastal waters, fish, pelagic mammals and birds has been widely reported. The subject has been reviewed extensively.[98, 100] Mercury is highly toxic to all forms of life, generally more so in the organic form than as an inorganic compound or as mercury vapour.[98] Acute poisoning can be fatal and otherwise causes permanent damage to the central nervous system. Such poisoning has been observed in exposed human populations in Japan, at Minamata and Nigata, as a direct result of the eating of contaminated sea foods.[59]

The World Health Organisation's limit for human food is 0·5 ppm and the Food & Agriculture Organisation's practical residue limit is 0·02–0·05 ppm. Some countries stipulate no residues (e.g. the USA). This is somewhat impractical, because fish collected from areas with no known source of pollution may contain up to 1·0 ppm (fresh weight) of mercury.[72]

Major sources of mercury include aqueous discharges from chlor-alkali plants, electrical industries, industrial catalytic processes, and paper manufacture; run-off from lands treated with organomercury fungicides; and from the mining and refining of mercury itself. There are many minor uses (e.g. in paints, dentistry). Mercury is rapidly scrubbed out of the atmosphere by precipitation, much finding its way into inland water courses or directly into the oceans.

The background (natural) sea concentration is between 0·03 μg/l and 0·3 μg/l.[88, 100] The methyl mercury constituent is probably about one-hundredth of this level. Turnover time is approximately ten years. A limited quantity of oceanic mercury is concentrated in bottom deposits, in contrast to river systems, where between 80 and 98 per cent may be in this form. Rivers contain less than 0·1 μg/l water if unpolluted, except in areas with natural mercury deposits. (In Minamata Bay levels of 1·3–3·6 μg/l were recorded during the periods when pollution was at its peak.) There is evidence to suggest that inorganic compounds can be naturally converted to the more toxic and soluble dimethyl mercury.[101]

Low levels ($<0.1\mu g/1$) of some organomercury compounds can inhibit photosynthesis in some species of plankton and affect fish behaviour. Normal levels in fish are about $0.01-0.2$ ppm. But, as with DDT, marine organisms can accumulate mercury to concentrations up to 500 times greater than ambient water levels. Fish can tolerate well over 1.0 ppm, with consequential health hazards to man, especially in communities dependent upon fish as a staple food. Accumulations have also been recorded in birds, especially amongst fish-eating species, and in terrestrial mammals. Foods other than fish contain mercury, and these must be accounted for when calculating total human dietary intakes. There is no evidence of any harmful effects to man from present levels of mercury in food in this country. Nevertheless because of its persistency in the environment every effort should be made to reduce all forms of mercury pollution.[102]

There are conflicting views on man's contribution to mercury in the environment. Estimates range from 1 to 50 per cent of the global load from human activities. Present discharges are estimated at 10^4 tons mercury per annum. Extensive monitoring programmes are now being established to observe mercury concentrations in various media, wildlife, foods and drink. Control of industrial discharges is improving, often with surprising savings of raw materials (for instance, losses of mercury from fifty US industrial plants have been reduced recently from 287 lb per day to 40 lb per day). The introduction of improved control techniques, better control of organomercury pesticides and technological advances obviating the need for mercury in certain industries will partially remedy the present situation. There remain, however, severe technical problems to be tackled in the mining and refining of mercury, and in industries where the use of mercury remains essential.

Lead demonstrates the complexities of studying multi-route access of pollutants to receptors. For instance, man inhales lead in gaseous and particulate form from vehicle exhausts and industrial emissions; ingests lead in food that contains lead residues and is externally contaminated by fall-out materials; and ingests lead in water and drinks contaminated by fall-out, industrial discharges and natural erosion.

Approximately three million tons of lead are produced annually, a considerable excess over natural supply. About 10 per cent of production is used in petroleum additives (e.g. tetra-ethyl lead).[95, 103] This is believed to introduce approximately 2×10^5 tons of lead to the oceans by rain-out, adsorption, precipitation and sedimentation. Only a small proportion is composed of highly toxic organo-lead compounds. Emissions from industrial activities—especially scrap-metal burning, car spraying, lead refining, ship building and breaking, wire making and organo-lead manufacture—and from the combustion of solid fuels contribute a further unknown amount to aerial loadings. The total consumption and the atmospheric emission of lead are both rising. This is corroborated possibly by reports of increased lead levels in arctic snow and ice samples. Lead in aqueous industrial effluents and from natural weathering may account for a further 2×10^5 tons annually. No estimate can be given for lead in solid wastes and sludges dumped at sea. Many near-shore sediments contain more lead than sub-surface sediments (older layers) in the same areas. The observed increase of lead pollution in the marine environment in general over recent years can only be viewed with concern.[95]

There is an extensive literature on the acute and chronic effects of lead upon man and animals.[21, 104] Despite the persistency, toxicity and accumulation of lead by some animals, there appears to be no evidence of damage to marine life or to man *directly* and *solely* from the ingestion of contaminated marine products. In urban areas, where blood lead levels are relatively high, marine foods rarely constitute a significant component of human diet. There are claims of incipient, widespread chronic lead poisoning in urban populations, especially among children.[105] Much overt lead poisoning in young juveniles has been identified as a result of ingesting leaded paints, a declining phenomenon in our cities as old dwellings are replaced. The association of mental abnormality in the very young with elevated blood lead levels depends on a number of factors, one of which is 'pica' (tendency to chew objects that may contain lead). Recently much attention and research effort have been devoted to lead from motor vehicles as another and perhaps more important factor in plumbism. Obviously children will absorb some lead from vehicle emissions, both by inhalation of atmospheric lead and by ingestion of food, etc, contaminated by fall-out. Other sources include discharges from local industry[106] and plumbosolvent townswater. The individual contributions of these various sources to total lead intake remain uncertain, although recent reports indicate that vehicle emissions are not a major source of lead.[10, 107]

The average UK human blood level is approximately $25 \pm 5 \mu g$ Pb per litre, with a considerable variation depending on occupation, residential area and diet. This average is close to levels quoted as unsafe for young children (e.g. 36 μg Pb/l) and as unsatisfactory for occupational exposures.[108] There is, therefore, little room for man-oeuvre with increases in exposure to lead from any source. Petroleum additives are particularly important in this respect because they are the growth sector of lead consumption, being automatically linked to increases in motor fuel consumption.[10,55] In various countries, includ-ing the UK, legislative and voluntary action to curb or even eliminate lead additives is being taken.

Other metals, including cadmium, nickel, chromium, zinc and arsenic, have been recorded in the atmosphere, rivers, oceans and various organisms. The routes of access to the seas vary, but direct discharges of industrial effluents are probably the major con-tributors.[95] Nevertheless aerial inputs are of undoubted significance in contributing to total environmental loads. Many metals are persis-tent and are accumulated to varying degrees, particularly in marine organisms in inshore and estuarine waters.

Cadmium is an important metallic pollutant, primarily because of its relatively high toxicity. It is accumulated by many animals, and is only slowly excreted by the human body. At low levels it is surprisingly toxic to fish under conditions of long-term exposure.[109] It has been associated with elevated rates of hypertension, heart disease and cancer in areas with high atmospheric concentrations and with soft townswater that dissolves zinc, cadmium and other metals from the piping in supply systems.[110] Wastes containing cadmium are gener-ated mainly from the extraction of cadmium itself and from the extrac-tion and refinement of zinc. Considerable quantities are also present in effluents from the paint and electrical industries (particularly the manufacture and reclamation of accumulators and batteries). In urban areas environmental contamination by cadmium may be ten to a hundred times greater than in rural areas, and much higher still around certain industries. However, there is little evidence that cad-mium is a truly global pollutant.

Nutrients
Nutrients include nitrogen, phosphorous, potassium and trace ele-ments such as silicon, manganese, sulphur and iron essential for the growth of plants and animals. The addition of nutrients to water, especially to standing waters, causes enrichment, or eutrophica-tion.[47,111,112] Eutrophication is essentially a natural process by which

waters receive nutrients by run-off from the land, by atmospheric fall-out and by deposition of vegetable and animal matter. Gradually lower forms of life (e.g. algae and plankton) dominate the ecosystem at the expense of free-living fish and other animals. This change is assisted by the reduction in light penetration below the surface and depletion of dissolved oxygen, particularly when higher plants die, eventually leading to anaerobic conditions. It can be shown that total productivity increases, although diversity of species decreases, as eutrophication proceeds.[111] The deposition of additional nutrients in industrial and domestic sewage effluents, fall-out of atmospheric pollutants and run-off from urban and fertilised agricultural lands greatly accelerates the whole process.

The major nutrients (by volume) are phosphorus and nitrogen, primarily in the form of phosphates, nitrates, nitrites and ammonia. Ammonia (average 5–50 μg NH_3/l sea water) is not persistent and is easily converted to nitrate. Similarly, nitrites do not present a serious problem in oceans. Nitrates and phosphates are soluble, persistent and readily absorbed on to particulate materials to form a reservoir of nutrients. Fortunately they are not concentrated by organisms, nor are they very toxic. The principal danger to the environment lies in their enrichment of fresh waters (chapter 2) and the ocean to produce phytoplanktonic blooms, followed by deoxygenation and anaerobic decay.[95] These processes destroy any other marine life in the immediate vicinity. Such phenomena have been observed in fjord and estuarine areas, are incipient in inland and brackish seas (e.g. the Baltic and Caspian),[95,103] and may occur occasionally elsewhere. Blooms of toxic organisms (e.g. dinoflagellates) may also be associated with nutrient enrichment. These blooms can be accumulated by filter feeders (e.g. edible shellfish) resulting in paralytic poisoning of humans.[118]

No attempt has yet been made to compute global marine loadings of nitrates and phosphates contributed by natural processes and human activities. Estimates have been made for the US[88,114] and for various OECD countries.[47] For instance, the total discharge of phosphates in the US was estimated at 436,000 metric tons of phosphorus in 1968.[88] It is inevitable that more nutrients will be produced as human populations and demand for food, etc, increase. Most of these additions will eventually reach inland waters and the seas unless effective measures are taken to reduce their consumption and to encourage nutrient recycling. In general, however, enrichment is essentially a composition of local problems in specific lakes, estuaries and enclosed seas where eutrophication is incipient or already serious. Under these circumstances unilateral local action would appear to offer the best prospects of control.[88]

Enrichment in estuarine and inshore areas is potentially serious in global terms because over 90 per cent of fish species use these areas as residence, passage and breeding zones.[95,115] (Thus one nation's wastes can affect future international food supplies.[116]) Also the destruction of most of the biological output of the oceans could occur *before* the seas are affected throughout by accelerated eutrophication (or by other forms of pollution). The control of nutrient discharges is economically and technically difficult. However, better management of consumption (e.g. of fertilisers and detergents) and waste disposal (e.g. of sewage effluents and sludges) can be effective locally in reducing the nutrient burden of rivers, estuaries and coastal waters.

Oil and associated materials

Oil creates global problems in its use and transport from widely dispersed sources.[60,101] The ubiquity of oil pollution has been amply demonstrated by technical and popular reports. Oil and oil products arrive in the sea directly or indirectly by discharge from sewers, land drains, industrial outfalls and spillage into rivers and seas; by submarine seepages at offshore oil fields; by shipping operations and accidents on the high seas and in ports, and by atmospheric fall-out (primarily of gaseous hydrocarbons).

The incidence of oil pollution from tankers and other ships is greatest along the main shipping lanes of the world. The primary routes of oil movement are along the continental shelves from the Middle East to Europe and Japan and from the Caribbean to Europe and North America.[60,101] Supplementary routes exist from New Guinea and are developing from offshore drilling sites in the North Sea and elsewhere. The continental shelves, including estuaries, are also the areas of greatest biomass. The risk of accidents[117] increases both with ship movements and with the narrowness of waterways (e.g. channels, entrances to ports and harbours). Total tanker cargoes are estimated at 650 million metric tons in 1967, 1,500 million in 1970 and 2,900 million in 1980.[103] At present about 5–10 per cent of the total polluting load may arise from tanker accidents.[88] Fall-out of airborne materials from fuel combustion may add a substantial quantity, particularly as hydrocarbons, to this basic load. The FAO estimates[103] that up to 1 million tons of oil come from marine operations and up to 5 million tons from river discharges. Estimates of the total discharge or basic load per annum vary between 1 million and 10 million tons.[91, 101] However, it is sobering to reflect that one serious accident to a modern super-tanker could increase the annual basic load by about 20 per cent.

The types of oil and petroleum products discharged vary enormously in their physical and chemical characteristics.[103] This results in great variations of toxicity, persistency and rates of breakdown, complicated by the prevailing environmental factors and the sensitivity to oil pollution of various receptors. Thus predictions of damage and dispersal based on one incident alone can lead to totally misleading estimates in different situations.

The effects of oil and oil products upon the environment include:

1 Despoiling of coastal amenities (e.g. oil on beaches).

2 Oiling of seabirds, which destroys water repellancy and prevents flying and may result in asphyxiation and poisoning.[60] (Oil fouling also decreases the resistence of birds to infection and other stresses.)

3 Poisoning of marine plants and animals, especially filter feeders in inter-tidal areas. Plankton may be adversely affected. The juvenile stages of some marine organisms are more sensitive than adults, resulting in long-term adverse consequences for the survival of the receptor species and its predators.

4 Indirect poisoning of marine biota by oil dispersants, etc, applied in remedial treatments. (Much attention has been paid to the development of non-toxic dispersants.) Mixtures of oil, water and detergent may be more toxic than the detergent alone. Sinking agents (e.g. sand) often merely settle on the bottom, creating more damaging situations than before. In general, oil spills should be left alone unless they are in enclosed or inland waters, or pose a serious hazard to water abstraction, amenity and wildlife.

5 Reduction of dissolved oxygen by bio-degradation of pollutants in enclosed waters can cause extensive mortality among local fauna and flora.

6 Tainting of commercial species of fish and shellfish (e.g. ≥ 0.01 per cent oil taints shellfish).

7 Alteration of animal behaviour affecting feeding, breeding and migratory responses of some animal species.

8   Some oils contain carcinogenic substances which, if they enter food chains and become concentrated, could pose a danger to species at higher trophic levels, including man. However, this source is only a very small contributor to the total human intake of carcinogens.

9   Long-term degradation of the marine environment, incorporating all the aesthetic and biological effects discussed above.

10 Enhancement by oil films of the absorption of atmospheric hydrocarbons and organochlorine compounds, thus increasing the rate of pollution of the surface layers of the ocean. This could accentuate the exposure of planktonic organisms to pesticides, etc, with detrimental results to plankton, their predators and the rest of the food web.[88]

Evidence of extensive, long-term damage to oceanic resources due to oil pollution is hard to find.[103] Apart from the aesthetic, bird damaging and local fouling problems, it is impossible to gauge the significance of oil pollution hazards. Fortunately public outcry against loss of amenity and damage to birds has led to extensive and continuing action to limit the risks of accidental oil spills and general discharges from tankers and other ships. The majority of the World's tanker operators have agreed to a maximum residue discharge rate of sixty litres per mile, at which rate oil is supposed to be dispersed and not to form a slick.[60] International agreements and industrial compensation systems seek to reduce and eventually to eliminate oil pollution on the high seas.[95] There are still difficulties to be overcome, however, in developing safe cleaning-up methods, in the enforcement of load-on-top procedures and in the provision of safer tanker routes and shipping control.[60,88]

Tankers, together with other marine operations (e.g. other ships, the loading and unloading of oil), contribute only about 15–20 per cent of the total oil burden.[95] Eighty to eighty-five per cent of oil is derived from discharges and run-off from factories, sewers, refineries, urban areas, garages and dumps going directly into the sea or *via* rivers and canals. Clearly there is a wide range of non-marine sources of oil that should be more closely controlled. There is, in addition, an unknown potential for pollution inherent in offshore drilling for and storage of oil.

The contribution of atmospheric hydrocarbons to the pollutant load of rivers and seas is a long-term problem that is closely related to the use of oil and petroleum as a fuel for powered vehicles.[88] Until

vehicular emissions are controlled this source will remain and probably increase in magnitude with any increase in vehicle fuel consumption.

CONCLUSION

Clearly it is difficult to determine from the limited data available the
relative importance of each pollutant and its global consequences.
However, the most critical factors influencing global climate in the
longer term may be particulate and aerosol emissions from all sources
(e.g. intensive land use, atmospheric emissions). In contrast, the most
serious threat to terrestrial, aquatic and marine ecosystems seems to
be intensive agriculture, with its mono-crop cultures, pesticides, fertilisers and waste disposal problems. Nutrients from all sources will
probably present serious problems in controlling the quality of aquatic and marine waters. At the same time the long-term environmental
hazards associated with any massive expansion of nuclear power production and offshore oil extraction cannot be ignored.

CHAPTER 4

The measurement of damage: objectives and methods

Damage studies have three main objectives:

1   The determination of thresholds (i.e. minimum combinations of pollutant concentration and duration of exposure) for total damage and for individual responses (figure 6),

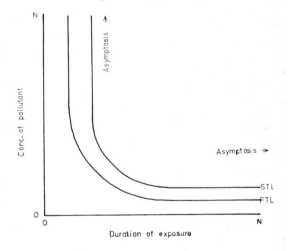

FIGURE 6   Typical asymptotic thresholds or time/concentration curves. *FTL* = first tolerance or threshold limit of maximum combinations of *c* and *t* which do not induce a response (level of 'no effect'). *STL* = second tolerance or threshold limit of minimum combinations of *c* and *t* which induce a response

2   The determination of dose/response relationships or functions (i.e. the way in which changes in concentration and duration of exposure alter the magnitude of response—figure 7);

3   The measurement of total damage suffered by a receptor in response to pollution under environmental conditions.

The nature and use of both thresholds and dose/response functions are discussed in some detail in chapter 5. Briefly, however, thresholds

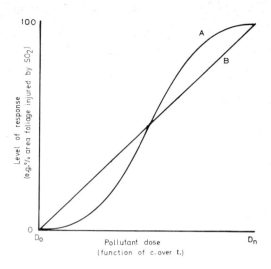

FIGURE 7   Relationship between dose and a single response: c = concentration of pollutant; t = duration of exposure; A = linear response; B = a non-linear response (sigmoid example)

provide the technical basis for environmental quality criteria in pollution control. Dose/response relationships have a similar function but may also be used in the rapid estimation of damage (chapter 6). Measurements of total damage are important information for pollution control procedures. But they must relate to populations, communities, national production, etc, rather than to individual receptors. Also, measurements should be in units suitable for economic evaluation.[19] Calculated economic losses can then be used to determine priorities for control and the benefits to be achieved by reducing levels of pollution.

Unfortunately we are woefully short of achieving all the above objectives (as has been demonstrated in the preceeding reviews of pollution and damage). This ignorance is due both to the sheer amount of work involved and to the inherent technical limitations of the methods used to measure damage—limitations which influence the quality of the damage data itself.

EXPERIMENTAL METHODS OF DAMAGE MEASUREMENT
Most damage data are derived from toxicity tests in which a pollutant of known concentration is delivered to an individual or group of receptors under controlled environmental conditions, usually in the

laboratory. Such tests are limited by the size of the receptor or populations of receptors that can be handled, and by the duration of exposure that is feasible without artificial aids.[118] In the case of humans, tests are restricted to exposures with only temporarily injurious effects.

The functions of toxicity tests are to identify the symptoms and effects of pollutants on receptors; to identify the toxic limits used as criteria in survey research and in environmental control; and to confirm field experience of pollution damage. The acute toxicity test has been by far the most significant contributor of data in these areas.

Acute toxicity tests

Receptors are exposed to relatively high concentrations of a pollutant for predetermined periods of time in a carefully controlled environment (e.g. constant temperature). The pollutant is usually supplied in the medium (i.e. food or water or air) either as a single dose[119] or at a continuous concentration throughout the experiment. Systems in which the medium is changed periodically or continuously are preferred because they maintain a reasonably constant environment and stabilise the pollutant concentration within the medium.[118] Such systems can achieve accuracies of less than ± 5 per cent variation in the level of response. The simplest expression of acute toxicity is the ranking of receptors in comparison with the response of a standard species (e.g. alfalfa as a sensitive standard for SO_2 injury[120]). This procedure gives no indication of threshold concentration or exposure time. Furthermore different exposure conditions change the rank order of many receptors.

For greater precision, the threshold concentration for a given response of predetermined magnitude may be discovered by exposure of receptors over a standard period of time. For instance, fish are often tested for $LC_{50/48hr}$ values, viz. the least concentration of the pollutant required to kill 50 per cent of a test population in forty-eight hours of exposure. Other expressions of toxicity or damage are used, but they are by no means fully compatible with each other. Thus the $LC_{50/48hr}$ value may not be interchanged with the median threshold limit value ($Tlm_{50/48hr}$) for the same receptor, because only the latter guarantees asymptosis (parallel progression) with the concentration (figure 8).

Threshold or tolerance limits are also used in air pollution tests with plants. Here the first (FTL) and second (STL) threshold limits

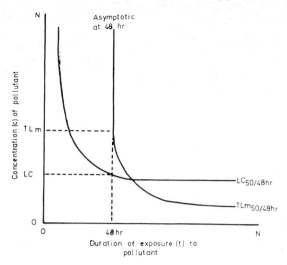

FIGURE 8   Comparison of two different measurements of toxicity of one pollutant to fish. $Tlm_{50/48hr}$ = threshold limit median value (concentration required to kill 50 per cent receptors becoming asymptotic after forty-eight hours' exposure). $LC_{50/48hr}$ = least concentration (required to kill 50 per cent receptors after forty-eight hours)

describe the combinations of concentration and duration of exposure required to cause no injury (maximum combinations) and first significant injury (minimum combinations) respectively (figure 6). These usually refer to foliar reactions, the STL being measured when 1–5 per cent of the foliage is visibly injured.[121]

Other terms include the median survival time for animals (duration of survival as a function of exposure to a predetermined concentration) and various dose and damage rates such as rad./year (radiation dose), mil./year (loss of metal thickness due to corrosion) and doubling dose (dose required to double a given rate of damage). Sometimes the estimated dose for a given damage rate (e.g. the ED_{50}) is calculated by interpolation or extrapolation of test data (e.g. from LC_5 and LC_{25} values.

The most surprising characteristic of toxicity testing is a general failure to determine threshold values over a sufficiently wide range of concentrations and durations of exposure to discover if or when asymptosis (figures 6 and 8) occurs along both axes. This applies not only to threshold or tolerance limits but also to other time/concentration functions throughout the range of magnitude (e.g. 0 to 100 per cent) of any given response.

Toxicity data from acute toxicity trials may be extrapolated to the external environment with confidence only where the conditions of exposure are strictly identical (e.g. same doses, temperature, receptor health). With each deviation from test conditions margins of error greatly increase. For instance, the threshold limit of SO_2 injury to plants can range between −50 per cent and +250 per cent of the test value, depending on the age, health and environmental circumstances (e.g. water availability) of the receptor.[122] Similarly, acute toxicity data for fish may seriously underestimate damage in the field (e.g. actual toxicity is 0·4–0·6 of anticipated $LC_{50/48hr}$ (depending on the pollutant involved) because of the greater sensitivity of some stages of the receptor (e.g. juveniles) and subtle interference with feeding and breeding).

There is no guarantee that (*a*) an order of relative toxicity of either pollutants or receptors (figure 9), (*b*) a type of response, or (*c*) the long-term consequences of pollution will remain the same under different conditions of exposure, especially chronic exposure. Such variations cast serious doubt upon the use of acute exposures in physiological research and in the interpretation of chronic damage.

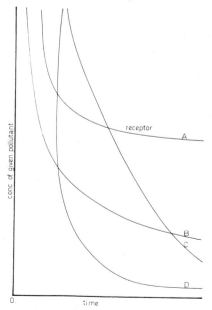

FIGURE 9   Relative toxicities of several pollutants to one receptor

The use of substitute receptors is unavoidable in studying the serious effects of pollution on humans. However, there are numerous toxicological and symptomological problems in such work, the most obvious being the differences in lifespan, sensitivity and response between proxy and subject.[123] Animal experiments should be regarded as *indicators* of possible damage in humans until similar responses are confirmed by direct experimentation and epidemiological experience.

Chronic toxicity tests

The techniques are essentially the same as those utilised for acute toxicity tests, although the equipment may be more elaborate to maintain prolonged periods of exposure. The receptors are exposed to relatively low concentrations of pollutants over longer periods of time. The results depend very much on the criterion of chronic toxicity adopted and on the duration and concentration of exposure applied. [28, 118, 123]

Frequently, chronic *lethal* toxicity tests are merely extensions of acute trials; they may extend knowledge of lethal time/concentration functions.[124] In some cases it is suspected that different mechanisms of toxicity may be involved in acute and chronic lethal exposures (e.g. the toxicity of cadmium to fish[124]). Greater attention is now being paid to the study of lethal and sub-lethal responses to very long-term exposures, even for periods equivalent to the life span of the receptor. Such tests are time-consuming. They present severe technical problems, especially in maintaining stable pollutant concentrations and receptor health (e.g. antibiotics may be needed to control disease). The limited data available indicate that the basic forms of thresholds and dose/response relationships (figures 6 and 7) are similar for responses to most types of exposure.

Besides morphological changes, exposure to pollution induces biochemical and physiological responses in animate receptors. For instance, low-level exposures of some grasses to SO_2 can result in considerable loss of productivity without obvious injury to foliage.[4] Exposure of humans to carbon monoxide increases the carboxy-haemoglobin content of the blood and has been claimed to be associated with heart disorders, headaches, loss of visual acuity and impairment of discriminatory functions.[125]

The accumulation of persistent pollutant residues within a receptor may be studied in chronic exposure tests. The basic objectives are to determine the level of residues within a receptor that constitutes a threshold for a damaging response, and the rate at which accumulations of residues occur in response to exposure. Studies may include the determination of concentration factors between media and receptor (e.g. 1 : 1 for mercury in sediments to fish[126]), lower trophic orders and receptor (e.g. 1 : 1·5 to 2 for DDD in plankton to herbivorous fish[127]), and between various organs, tissues and fluids of the receptor. The last factors are especially important where residues are stored harmlessly in particular tissues (e.g. DDT in fat) but subject to release into the bloodstream with toxic consequences during periods of stress (e.g. starvation). They are also important in radiobiology, where the different sensitivities of organs necessitate the determination of organ as well as whole body exposures to radiation.[127]

The determination of levels of pollutant and residues associated with significant damage (e.g. the death of fish where DDT exceeds one part in 10^9 water and 10^6 muscle tissue[128]) is difficult, especially when dealing with subtle responses (e.g. egg shell thinning in birds[68]). Toxicity tests rarely duplicate the stresses and exposure conditions that are encountered in the field. Body fluids (e.g. blood) may be suitable media for testing exposure to pollution (e.g. to lead), since they do not involve destruction of the receptor during sampling procedures. But they do not necessarily reflect longer-term accumulation of residues in other tissues and organs.

Chronic toxicity testing of sub-lethal responses has three related uses. Firstly, it assists the detection of less obvious adverse responses with long-term damaging consequences for the receptor. Secondly, measurements of residues in fluids have an indicative role where they can be related to tissue, etc, accumulations of residues and other adverse responses for which threshold limits have been determined. Thirdly, sub-lethal responses not only exhibit dose/response relationships but also form part of a composite relationship in which dose levels are associated with different responses (chapter 5). These two types of function can be seen in figures 6 and 10 respectively. Fluid residues (e.g. of fluorine, lead, carbon monoxide) can be substituted for dose levels (figure 11), thus serving as indicators of damage.[129]

Variations in receptor age or stage of development, environmental conditions, the duration and concentration of exposure, and the numbers and types of pollutants all affect the nature of the response of a receptor (phenomena discussed in chapter 5). Systematic studies of such variables using acute and especially chronic toxicity testing

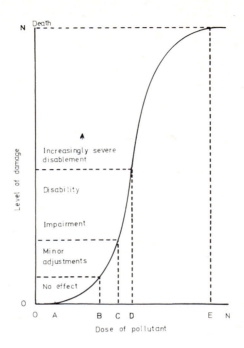

FIGURE 1ɔ   Theoretical relationship between dose and various responses or degrees of damage

Dose

O to A = Adjustments within normal range (homeostasis) of receptor. No effect.

A to B = Physiological, etc, adjustments (e.g. elevated levels of lead in blood) but no significant cost to receptor except over very long periods (e.g. lifetime) of exposure.

B to C = Some impairment of function. Short-term subtle effects on behaviour, etc, with possible long-term effects upon mortality.

C to D = Some disablement (e.g. injury of foliage, reduced growth). Not permanent in shorter term but serious (e.g. osteosclerosis) over long periods.

D to E = Increasingly permanent disablement, advanced morbidity and mortality.

Above E = Death.

techniques are not common.[124, 130] Such studies are essential if toxicity tests are to evolve into reliable methods of forecasting damage in the field. Better synthesis of field conditions and closer integration with field experiments and surveys are also essential.

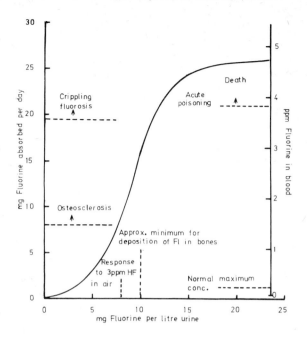

FIGURE 11   Relationships between fluorine absorbed, fluorine content of two body
fluids and some responses to fluorine poisoning[129]

Experimentation in the field

Experimental field research into pollution damage is a logical exten-
sion of laboratory techniques at one extreme and a controlled form of
survey research at the other. The degree of control of external variables differs with the approach adopted. At the 'laboratory' end of
the spectrum a receptor may be exposed to a pollutant introduced
into a previously unpolluted environment. Alternatively, healthy
receptors may be transferred (e.g. lichen transplants) to an environment in which exposure to pollution, sunlight, temperature, etc, are
uncontrolled but are carefully monitored. A closer approximation of
laboratory conditions is achieved by transferring the polluted medium
from the environment to the laboratory, which permits some stan-
dardisation (e.g. pH adjustments[131]).

   All these approaches attempt to simulate the field situation, preferably with maximum control and minimum interference. Frequently,
however, they demonstrate the considerable differences in response
that occur between exposures in the field and in the laboratory.

Again, the factors creating such differences must be determined if reliable, predictive field techniques are to be evolved.

Much field work falls between field experimentation and survey studies. Experimental activities are illustrated by the use of radio-tracer techniques, modelling studies of polluted ecosystems, and analyses of pollutant residues and damage in 'transplanted' receptors. Survey techniques are, however, widely used in sampling media for pollution and receptors for residues[10] and for damage (including changes in communities and ecosystems[132]).

Quantitative information about damage in the field is extremely sparse because of methodological difficulties,[133] particularly in the selection of a suitable index of damage. Satisfactory indices of productivity loss[134] have yet to be devised. In most polluted systems there is also some loss of diversity owing to the different sensitivity of receptors to pollution. Unfortunately indices of productivity and diversity do not always relate to changes in pollution in the same way; both may decrease at proportionately similar or different rates with increasing pollution. In the case of eutrophication productivity can increase while diversity decreases.[111] It has often proved easier to use biological indicators of both pollution and damage (e.g. lichens and SO_2, trout and water pollutants, species of algae and eutrophication). These offer a possible quantitative measurement of the status of the medium, community or ecosystem.

SURVEY METHODS OF DAMAGE MEASUREMENT

Epidemiology, the science and study of epidemics, is used in medicine and plant pathology to embrace all surveys and allied studies of the incidence and severity of diseases and disorders, including those caused by pollution. The following discussion leans heavily upon medical science. Nevertheless there are many similarities between all field surveys of pollution damage to animate and inanimate objects. It is a fundamental rule of such studies that a pollutant remains a speculative agent of damage until the following criteria of causation are satisfied:

1   Damage is associated with the pollutant in time and in space (i.e. it is not a unique, isolated phenomenon).
2   The incidence and severity of damage are correlated with a range of concentrations and durations of exposure (i.e. reveal time/concentration and dose/response functions—chapter 5).

3   Such correlations and relationships are consistent under similar conditions of exposure.
4   Damage is specifically and uniquely related to the one pollutant but not to any other agent of damage.
5   Epidemiological experience is consistent with experimental evidence.

Often the last two criteria cannot be fully satisfied. The majority of responses are non-specific (e.g. chronic bronchitis), being associated with more than one causal agent (e.g. coal smoke, SO_2, cigarette smoking). Consistency between experimental and epidemiological experience is difficult to achieve because of non-specificity of response and the technical difficulties of duplicating field conditions in the laboratory. Specificity may, however, be conferred by secondary tests (e.g. analyses of lead in blood and tissues). Disagreement does not, therefore, guarantee that either type of data is false. But any agreement between such independent sources inevitably increases confidence in the result.

Two axioms of statistics are particularly relevant to the analysis and interpretation of survey data:
1   Irrespective of the method of statistical analysis used, the results are limited in accuracy by the quality of the data input.
2   Statistical correlation between exposure to pollution and level of damage indicates a possible causal relationship; in the absence of experimental evidence it cannot prove the existence of damage.

All these points should be remembered when considering survey data. The quality of the data input deserves particular attention, especially information obtained by monitoring of the environment.

Monitoring
Monitoring has been defined as the systematic observation of all parameters relating to a given environmental situation in a manner designed to establish the characteristics of those parameters and their changes with time.[88] In pollution studies there are four main categories of relevant parameters which may be monitored:
1   *Pollutants*: concentration and duration of exposure for each type of pollutant (monitoring of the environment in relation to pollution in the UK has recently been reviewed[14]).
2   *Non-pollutant environmental parameters*: characteristics of medium (e.g. flow rate, pH), weather, climate, geology, etc.
3   *Receptors*: receptor types and their distributions.

4 *Damage*: level and/or numbers of receptors damaged in (3),
   which may be expressed as a damage rate (e.g. x deaths/1,000
   population) or as a crude loss (e g. y deaths).

Within each category secondary or ancillary parameters may prove
important in the analysis of damage. For instance, in human studies it
is useful to know the socio-economic status of receptors, their age/sex
ratios, occupations and habits. In most cases it is also desirable to
establish the baseline characteristics of the environment (e.g. pH),
receptors (distribution, etc) and responses (e.g. death and morbidity
due to non-specific diseases) in similar, unpolluted or 'controlled'
areas. Natural variations among receptors over long periods of time
are not, however, easy to determine.

Nevertheless the correct selection of control populations is
extremely important in experimental and survey studies. Ideally they
should be identical to the exposed population in every possible way
apart from exposure and response to pollution. Furthermore where
sampling of receptors is undertaken random sampling procedures
should be used to select individuals for examination. (There are limitations to this requirement in certain types of research.)

The methods and technical problems of monitoring pollution,
environmental parameters and receptors are dealt with in detail in
various specialist publications.[14, 29, 40, 135] Chapters 2 and 3 briefly indicated the information generated by existing monitoring systems.
However, greater use could be made of modern techniques to
improve knowledge about the distribution of pollution damage, especially to human health, plant life, materials, fisheries, wildlife and
local amenities. For instance, remote sensing or surveillance (e.g. by
infra-red aerial photography) has proved helpful in providing preliminary data of the dispersion of pollution, and its effects upon
receptors, especially plants.[136] Biological monitoring of pollution and
damage is also frequently advocated.[14]

However, there are certain considerations to be borne in mind
when devising a biological monitoring system:[137]

1 It must reflect the pollutant levels and environmental circumstances under investigation (e.g. lichens are unsuitable for studying diurnal variations in SO_2 levels).
2 It does not significantly deplete the indicator population.
3 The indicator is accessible, responds to a wide range of exposure
   conditions without being destroyed, and reflects pollution trends
   within its related community and ecosystem.

4 The system is the most practical in terms of consistency of results, coverage of exposed area, and cost (including efficiency in manpower, etc).

Biological monitoring is useful in revealing accumulations of residues (e.g. pesticides in birds,[68] radio-nuclides in marine biota[75]). The geographical distribution of pollution can also be reflected in the pattern of survival of an indicator species (e.g. lichens in response to SO_2[138]). However, biological monitoring is a technique which indicates recent but nonetheless past levels of pollution. Furthermore it is insensitive to short-term variations in exposure. Biological monitoring is therefore best suited to the study of long-term trends in pollution damage to receptor populations, communities and ecosystems.

Existing technical monitoring systems deal mainly with pollution. They have been evolved mainly to cope with local pollution control problems. The result is a diversity of systems which are not always compatible with the requirements of damage studies. Comprehensive systems for monitoring all environmental parameters throughout the country would be extremely costly and probably unnecessary for most control purposes. An alternative might be to establish a skeleton system of sampling stations supplemented by local intensive systems for monitoring particular areas of extreme and 'representative' geographical and pollution conditions. Such a system is being developed for the monitoring of airborne heavy metals. The general inadequacy of existing systems in all media is highlighted in a recent report which proposes extensive reorganisation of pollution monitoring in the UK.[14]

Types of survey

There are four basic types of damage survey: (*a*) observational, (*b*) retrospective, (*c*) present (also called one-time or one-off surveys), (*d*) prospective.

Observational surveys These are confined to initial studies of acute, episodic pollution (e.g. London smogs, the *Torrey Canyon* oil spill[139]). Damage is usually considerable in extent and severity. Operations tend to be *ad hoc* and to rely upon fortuitous measurements of pollution and damage by monitoring systems already operating in the polluted area. The results are often qualitative and indicative. The data are therefore unsuitable for extrapolation or interpolation to different situations involving the same pollutant, especially those involving low-level exposures.

Episodic events serve to stimulate public opinion and political action, and to generate more systematic research. In fact the border-lines between observational, one-time and retrospective surveys are indistinct where the latter are applied as 'follow-up' studies of episodic pollution damage.

Retrospective surveys   These[17] rely on fortuitous measurements of past pollution and damage, but they may include measurements of current damage associated with past exposures to episodic (acute) or chronic pollution. The information (e.g. monitoring) systems are not designed with such surveys in mind, and the results are therefore subject to considerable error. Two quite distinct approaches may be recognised:

1 Analysis of institutional records of pollution levels, receptor status, etc, and damage (e.g. human mortality[17,140]) which attempts to correlate pollution with response. Where sufficient variations occur in exposure and in response rates multiple regression tech-niques may be applied directly, without any selection of 'unex-posed' control cohorts of receptors for comparative purposes. The margins of error can be considerable, because data sources are uncontrolled and important variables (e.g. other pollutants, smoking habits in humans) are not included in institutional records.

2 Study of receptor cohorts from both exposed and unexposed (control) populations to determine the effects of past exposure to episodic or chronic pollution. Again there is great reliance upon fortuitous measurements of pollution and damage, especially from institutional sources. Receptors may be examined in detail for damage and other important variables such as socio-economic status, smoking habits, occupational exposures and habitat. How-ever, the relatively small populations that can be handled and the defection of some severely affected receptors limit the accuracy of the final result.

There are many variants of these types of survey which may over-lap with one-time techniques. For instance, surveys of oiled, dead seabirds are essentially retrospective studies of episodic pollution but they also indicate the current damage inflicted upon the bird popu-lation. Irrespective of the type of retrospective survey conducted, discrimination between cause and effect becomes increasingly difficult as pollution decreases in intensity and other causal factors become more influential in determining the nature of responses.

Generally, retrospective techniques are best suited to the study of well defined responses to episodic pollution.

Present-time (one-time, one-off) surveys   Retrospective data are often used as bases for these surveys and for experimental studies. Present-time surveys—studies of responses at specific points in time (analogous to public opinion polls)—are widely used in pollution damage research. Unlike previous techniques, there is considerable scope for monitoring pollution, receptor populations and other relevant factors. A pollution monitoring system may be designed specifically for a particular survey.

Repetition at discrete intervals updates information and permits the determination of trends in pollution and damage. Most studies utilise random sampling techniques for statistical accuracy. With marking techniques (e.g. tagging), population changes, survival rates, etc, may be studied. There are, however, disadvantages, in that symptom definition and follow-up studies may be hindered. Repeated sampling of the same individuals each time overcomes these difficulties but creates additional problems when allowing for natural wastage (deaths, emigration), recruitments (immigration) and damage during collection of samples.[141]

Much field information about pollution damage comes from present-time surveys conducted in isolation or in series. Costs and logistics, however, tend to restrict their size in terms of both numbers of receptors and number variables included. Nevertheless such survey techniques are well suited to the study of acute and chronic pollution damage provided the responses involved are reasonably well defined.

Prospective surveys   These are the most sophisticated techniques of epidemiological research.[140, 142] Usually, random cohorts of receptors from unexposed and exposed populations of receptors are studied from birth (installation in the case of materials), or from some common state of health, on into the future until death or destruction. Successive generations may be studied where pollution has long-term genetic effects (e.g. radiation damage). Monitoring services can be tailor-made for the operation.

Prospective surveys are undoubtedly the best method of surveying the long-term subtle effects of chronic, low-level environmental contamination, provided that the symptoms of damage are detectable and measurable. One possible application of prospective techniques might be in an investigation of juvenile sensitivity (e.g. bronchitis[143]) to air pollution and its relationship with adult morbidity and mortality due to chronic bronchitis.[16] Prospective surveys are used infrequently in pollution research, however, because they are extremely costly. Workers are reluctant to commit themselves to such long-term research because of personal career considerations. Furthermore to maintain continuity of results it is essential to adhere to the original methods and information systems throughout the survey period. This raises problems of obsolescence and the compatability of data with information from more recent studies using improved research techniques. Paradoxically revision and the updating of methods lead to the phasing and discontinuity of results over time despite ensuring greater compatability with external data.

Interpretation of survey results
This section briefly discusses some problems associated with primary data and the interpretation of survey results.

Care must always be taken to determine those measurements of pollution most closely associated with damage. For instance, fish kills may be more closely correlated with peak daily pollutant concentrations in a river than with monthly or yearly average values. Frequently one pollutant is measured as a representative of other pollutants (e.g. SO_2 concentrations reflect those of smoke). However, changes in the relative levels of these pollutants may not be the same over time or in space.[10,55] This also applies to consumption (e.g. of fuel) and waste production indices used as proxies for pollutants.[55] Similar problems arise where a pollutant enters a receptor in several ways (e.g. lead in air, water, food);[76,127] changes in one input may be assumed to be representative of changes in total intake and exposure. The relationships between these various measurements of pollution should be determined before observed damage is attributed to a particular pollutant dose received by the receptors.

Several facts distort final estimates of damage. For instance, captive sub-populations (e.g. the sick) and minority groups of receptors should be avoided because they are atypical of the general population. Exceptions are usually confined to sub-groups defined by characteristics such as age (e.g. lead poisoning in young children) and sex (e.g. bronchitis in middle-aged males) where there is a particularly high risk of damage. Similarly, environmental factors can significantly alter the response of receptors to pollution (e.g. water availability affects the sensitivity of plants to SO_2 injury). This can be a very complex problem. For instance, latitude, rainfall, socio-economic status, population density, personal habits (smoking) and occupation can all be associated statistically with total and individual causes of human morbidity and mortality.[17,18,37] Some variables (e.g. socio-economic status) are particularly troublesome, having multiple associations with a range of parameters (e.g. occupation, population density, pollution levels, damage). However, their exclusion from surveys and experimental studies frequently results in excessive estimates of pollution damage.

Most errors in damage estimation result from poor sampling technique, the inherent limitations of survey techniques, and especially from the basic inadequacies of the input data. There is relatively little abuse of the criteria of causation, although overemphasis of certain aspects (e.g. pollution as dominant factor in the causation of lung cancer) and of maximum rather than middle-range estimates of damage are not uncommon.

The association of current pollution with current damage[18,140] has the merit of simplicity of interpretation. It is, however, misleading.

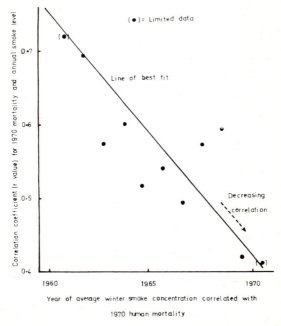

FIGURE 12 Relationship between 'current' total adjusted human mortality (1970) and air pollution levels in 'current' and preceding years (1960–70)[37]

Current damage is a product of past and present exposures to pollution, especially under chronic conditions. Furthermore such a simple approach can never distinguish between the exacerbation and the initiation of a disorder (e.g. chronic bronchitis). There is, therefore, a danger of overestimating the significance of current pollution levels. There are several possible approaches to this problem, provided the requisite data are available. For instance, linear regressions of current damage with annual average pollutant concentrations can sometimes (figure 12) demonstrate increasing correlation of current damage with each preceeding year's exposure.[37]

Similar problems arise with the equally common practice of assuming a linear relationship between dose and response which may appear so appropriate for the analysis of much survey data. However, detailed experimental and survey studies will often show that this relationship is non-linear (chapter 5). The effects of chronic exposures may be overestimated in such cases (figures 1 and 4).[144] The converse situation arises in the recovery of receptors from pollution, where it may be assumed that reductions in pollution produce the maximum benefits to the receptor instantaneously. Detailed studies have shown, however, that the rate of recovery over time is again non-linear (chapter 5).[145, 146] In fact recovered receptor populations may never quite achieve identical status to unexposed populations. Thus the long-term effects of pollution are understated and the benefits of stricter pollution control may be overestimated.

In conclusion, it is hardly surprising that surveys often fail to determine accurately the amount of pollution damage inflicted upon receptor populations. Poor data, the exclusion of critical variables, multicolinearity and oversimplification of dose/response relationships hamper interpretation and obscure the true level of damage.

Experimental studies are limited in their ability to recreate and to predict conditions in the external environment. Greater effort is needed to bring experimental and field experience closer together. Surveys deserve better experimental support facilities so that both approaches utilise the same basic material, thus promoting the interpretation of damage. Generally, greater emphasis should be applied to the study of chronic pollution damage.

CHAPTER 5

The estimation of damage: responses to pollution

This chapter considers in greater detail the characteristics of thresholds and dose/response relationships.

THRESHOLDS

The term 'threshold' has a variety of meanings. For instance, in medicine it usually refers to a minimum lethal exposure to poison. It has, however, a wider meaning as the concentration at or above which exposure for a given period of time will produce a particular response in a receptor. It is theoretically possible to determine thresholds for each such response in a receptor (e.g. residues in blood, loss of co-ordination, impaired lung function, death). The precision with which these thresholds are determined affects their value as parameters in damage estimation (chapter 6) and pollution control.

Exposure to a set concentration of pollutant over a given period of time permits the grouping or ranking of receptors according to their relative sensitivities. Table 3 gives examples, including one in which

TABLE 3   Examples of receptors, ranked according to their relative sensitivity to single exposures to a pollutant[120, 147]

(a)   Receptors ranked by comparison with reference receptor (alfalfa); sensitivity to SO$_2$ decreases with increasing index number

Alfalfa	1	Cauliflower	1·6	Potato	3·0
Barley	1	Apple	1·8	Onion	3·8
Spinach	1·2	Peas	2·1	Celery	6·4
Clover	1·4	Plum	2·5	Privet	16·0

(b)   Receptors ranked in groups according to concentrations sufficient to cause foliar injury[28, 147]

Group	Receptors	μg SO$_2$/m^3 air/24hr
I	Clover-type plants	430– 575
II	Wheat, leafy vegetables (excluding cabbage), beans, roses	575– 860
III	Roots, rape, cabbage	860–1,145
IV	Celery, privet, plane	1,145–2,000

different receptors are ranked by comparison with the response of a sensitive reference receptor. However, the order of sensitivity can change considerably with changes in environmental conditions, the age and health of receptors and especially with the concentration and duration of exposure to pollution. This means that simple ranking trials have little value in the measurement of damage except where exposure conditions in the field are identical to those in the trials.

Some refinement is achieved by determining threshold concentrations for a given level of response by exposing receptors over a range of time periods. This approach can be used to compare either the responses of one receptor to several pollutants or the response of several receptors to one pollutant (figures 9 and 13). The application of a range of concentrations and durations of exposure will indicate changes in the order of relative sensitivity.[147]

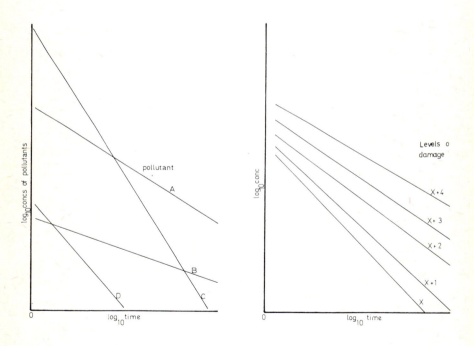

FIGURE 13   Linear conversion of threshold data (*a*) for one receptor and several pollutants (see figure 11), (*b*) for different levels of damage to a receptor caused by one pollutant.

Most threshold concentration data relate to acute exposures causing overt responses (e.g. foliar injury in plants, the death of fish). The determination of thresholds for more subtle responses, especially to low concentrations over long periods of time, is technically difficult. Nevertheless some have been determined (e.g. noise at or above 90 dBA is associated with psychological disturbance in man). An alternative approach is the interpolation or extrapolation of acute threshold values (e.g. 0·02 Tlm_{50} for zinc and copper induces avoidance reactions in salmon), the dangers of which have already been mentioned (chapter 4).[118]

The threshold values discussed so far are arbitrarily selected parts of a complex relationship between concentration (c) and duration (t) of exposure for a response of predetermined magnitude (e.g. 1 per cent foliar injury[121]). This relationship, sometimes called the threshold limit or function (TL), can be determined only by extended experimentation and observations of a particular response over wide ranges of concentration and exposure time. Consequently data are limited and relate mostly to acute exposures. Nevertheless studies with diverse receptors and pollutants indicate that the relationship between concentration and exposure time is non-linear (figure 13; chapters 1 and 4).

Workers studying the effects on plants of acute exposure to gaseous air pollutants[6,45,148] attempted to determine the first (FTL) and second threshold (STL) limits for no injury and the minimum level of significant injury respectively (chapter 4). Initially concentration and time were assumed to be reciprocal functions (table 4). Later it was found that changes in concentration were relatively more important than changes in exposure time in determining the level of injury. Formulae (table 4) were derived with constraints to allow for this anomaly and then elaborated as exponential functions to allow for asymptosis along both time and concentration axes (figure 6).

There are two major difficulties with such complex expressions. Firstly, they are specific to individual receptors, pollutants and environmental conditions, thus reducing their value as generalised functions for predicting damage. Secondly, there are anomalies such as pollutants exhibiting no asymptotic response (e.g. acryonitrile in fish toxicity trials).[124] Less elaborate but workable expressions of threshold limits are desirable.[8]

Simple graphical representations may be prepared by linear regression analysis or simply by hand plotting a curve to data available (e.g.

TABLE 4   Some mathematical expressions of the threshold limit[6,45,148]

(a)   Simple, reciprocal function for c and t

$$(C - C_R)t = K$$

(b)   Slight modification allowing for changes in c having greater effect than changes in t

$$C = K\left(\frac{1}{t}\right) + a$$

(c)   Exponential function

$$t = Ke^{-a}(C - C_R)$$

(d)   Linear logarithmic function

$$C \log_{10} = m - nt \log_{10}$$

where C = concentration of pollutant during exposure

t = duration time of exposure

C_R = asymptotic threshold concentration for injury

K = constant

a = constant to allow for dominance of c over t

e = exponential function

m and n = coefficients of regression

fish toxicity data[115]). With many pollutant/receptor interactions, how-ever, a more accurate linear function may be achieved by regression analysis of the logarithms of both concentration and time (figure 14).[45] Care must be taken not to include data from points beyond the onset of asymptosis along either axis.

The slope and position of the linear function for a response of given magnitude (e.g. 5 per cent injury) are specific to each pollutant/recep-tor interaction. A series of linear functions may be postulated for each level of a particular response of a receptor (e.g. 0 or FTL, 1–5 per cent or STL, 10 per cent, 20 per cent . . . 100 per cent). However, it is rarely possible to identify a complete set of functions for a given response, mainly because of the time and effort that are required for their determination. One finds instead that particular functions have been determined in different fields of pollution research (e.g. LC$_{50}$ or

Tlm₅₀ values for fish, second threshold limits for plants). Most such functions conform to the linear logarithmic relationship between concentration and duration of exposure up to the onset of asymptosis. Some examples are given in table 5. The more limited data of subtle responses (e.g. lung irritation by air pollutants) follow much the same pattern. There are exceptions to this pattern, most of which involve pollutants not exhibiting asymptosis (e.g. acryonitrile) or with more than one possible mode of toxicity (e.g. cadmium).[124, 149]

TABLE 5   Threshold limits for injury to alfalfa due to acute exposure to sulphur dioxide [45]

(a) *Level of visible foliar injury (%)*	(b) *Threshold limit*
0·1–1·9	$c \log_{10} = 3·16 - 1·70t \log_{10}$
2–10	$c \quad\quad = 3·26 - 1·73t$
11–25	$c \quad\quad = 3·19 - 1·56t$
26–50	$c \quad\quad = 3·37 - 1·49t$
51–75	$c \quad\quad = 3·67 - 1·35t$

$c = \text{ppm} \times 10$
$t = \text{hr} \times 10$

The threshold limits for each level of a given response are components of a further, more complex relationship which expresses the magnitude of response as a function of pollutant dose.

DOSE/RESPONSE RELATIONSHIPS

Definition of dose
A dose may be defined as the amount or concentration of pollutant delivered to the receptor over a period of time. It has been expressed in several ways, e.g.

$$\frac{c}{t^k} \quad \frac{c^k}{t} \quad t \times c \quad t \log_{10} + c \log_{10} \quad \frac{c - c_r}{t}$$

However, only those notations that express total contributions of (c) and (t) are satisfactory. Most others represent dose rates rather than total doses.

Dose/single response relationships
If concentration is held constant and exposure time varied or vice versa, then the magnitude of response is a non-linear function of the dose delivered. There are several possible functions, but sigmoid or quadratic relationships are most favoured.[144,150,151] A linear relationship can be established, however, by regressions between level of response and logarithmic transformations of dose (e.g. $c \log_{10}$ + $t \log_{10}$). Examples in table 6 include injury, damage (e.g. yield loss) and residue accumulation in some receptors. In a few cases a better correlation may be achieved by a further logarithmic transformation; this time, of response magnitude. Most data refer to acute exposures and overt damage. Similar dose/response relationships may be found in the effects of air pollution on materials, animals and plants; pesticides and PCBs on animals;[68] radiation and pesticides on humans;[52] and water pollutants on animals.

TABLE 6   Some examples of linear dose/single response functions[45,etc]

	Receptor		Dose/response functions (d = dose)
(a)	Sensitive tobacco plants	I(% foliar injury)	$= 2.14 + 0.0055d$ (log$_{10}$ pp100 m ozone + log$_{10}$ hr × 10)
(b)	Gladioli (sensitive cultivars)	Y(% yield loss)	$= 3.06 + 0.029d$ (log$_{10}$ pp 1,000 m HF + log$_{10}$ hr)
(c)	Various wild plants	N(log$_{10}$ number of plants)	$= 2.52 - 0.53d$ (log$_{10}$ g magnesite dust/m³ air/24 hr)
(d)	Gladioli (sensitive cultivars)	R(ppm F dry weight residue)	$= 2.98 + 0.0094d$ (log$_{10}$ pp 1,000 m HF + log$_{10}$ hr)
(e)	*Pinus* sp	R(pp 10 m S dry weight residue)	$= 1.06 + 0.021d$ (log$_{10}$ ppm + log$_{10}$ hr)

Not unexpectedly, investigations into the recovery of receptors following the abatement of pollution have demonstrated non-linear relationships between time since abatement and response (table 7). Most of these appear to be exponential functions, but logarithmic transformations provide reasonable linear approximations of the relationship between time and response.

TABLE 7   Some examples of linear relationships between response and time since abatement of pollution[45, 119]

Receptor	Response function
(a) Salmon exposed to DDT	R (log$_{10}$ ppm total DDT residues) $= 0 \cdot 285 - 0 \cdot 864t$ (log$_{10}$ yr since abatement)
(c) Humans exposed to cigarette smoke	D (log$_{10}$ deaths/1000 population) $= 0 \cdot 24 - 0 \cdot 69t$ (log$_{10}$ yr since abatement)
(d) Trout exposed to Dieldrin	R (log$_{10}$ mg Dieldrin/kg trout muscle) $= 1 \cdot 50 - 1 \cdot 35t$ (log$_{10}$ days since abatement)

Dose/multiple response relationships

The occurrence of several responses to a pollutant within one receptor has already been mentioned (chapter 2). A dose/response curve for fluorine poisoning was used as an example of a multi-response function indicating the various doses at which different responses occur (figure 11). Although some responses reveal themselves only at relatively high dose rates, others may be apparent throughout almost the entire dose range. The critical points are those doses at which 'no effect' is recorded for individual responses.[118]

There are two major problems in such work. Firstly, there is considerable variation in the responses of individual receptors to given doses of pollution and also in the relationships between symptoms and residues. For instance, severe symptoms of lead poisoning in humans have been associated with 83·94 μg Pb/110 g blood, with a standard deviation of 44·14 μg Pb/100 g blood.[153] Similiar problems have been encountered with fluorine[129] and carbon monoxide poisoning in animals. Secondly, the rates of adjustment amongst different responses to changes in dose rate are variable. Thus changes in amino-laevulinic acid and in urinary lead levels occurred after periods of one to two weeks and over three weeks respectively following exposure to lead pollution.

The relationships between different responses to a pollutant vary in form, although many appear to be linear arithmetic functions. Thus the yield loss of plants is directly proportional to the area of foliage suffering *acute* injury from exposures to gaseous pollutants. Similarly, loss of milk yield in cattle is closely associated with the fluorine content of their urine. Residues, however, are complicated responses to interpret. For instance, sulphates in vegetation may exhibit no quantitative relationship with injury in plants exposed to gaseous sulphur dioxide, whereas in others the relationship is a linear arithmetic function. On the other hand, damage may be logarithmically related to levels of residues (e.g. eggshell thinning and pesticide residues in birds[68]).

If responses such as residue accumulations and the blood content of pollutant are used as indicators of subtle damage or as environmental quality criteria it is essential that the relationships between dose and each response, and between individual responses, are clearly defined, together with their typical margins of error.

EFFECTS OF VARIATIONS IN EXPOSURE CONDITIONS
So far it has been assumed that the exposure conditions remain constant over time. Even when discussing changes in dose rate it has been assumed that the changes are from one steady state to another. However, conditions in the external environment are far from stable. This applies not only to concentrations and doses but also to other influential external variables such as humidity, pH and temperature.

Variations in the concentration and duration of exposure
A reasonable working hypothesis is that the effect of exposing a receptor to fluctuating pollutant concentrations, each of different duration, is approximately the same as the effect of constant exposure to the mean concentration of the pollutant over the same time period. This appears to apply in many acute toxicity trials,[131] provided that (*a*) the periods of low concentration are not long enough to permit receptor recovery (i.e. are below threshold limit for response) and (*b*) the periods of high concentration do not generate different and more severe responses from those experienced during the rest of the exposure. Thus damage should be associated only with those short exposures to peak concentrations which exceed the receptor's threshold limits (STL). However, fluctuations about the threshold limit can have a greater effect than constant exposure to the mean

concentration of the pollutant over the same period of time.[154]

Complex formulae have been derived to explain the effects of such variable exposures. An example is given in table 8 for injury to plants resulting from sulphur dioxide pollution exceeding the second threshold limit once only. More complex functions are required for multiple fluctuations about the STL to allow for increased stress, sub-threshold recovery periods, etc. The obvious drawback of such mathematical expressions is their extreme specificity to individual receptors; they can be used effectively for predicting damage in the field only after much experimental work with the receptor to determine its basic response characteristics.

TABLE 8   Effects of fluctuation about the threshold limit in plants exposed to sulphur dioxide, as developed by Zahn[154]

$$S = SR + \frac{7 \cdot 3}{K_2 P_2} \sqrt{c \, \log_{10} \frac{ti}{tR}}$$

where S = anticipated injury (0–5 scale)
SR = 1–5% injury, equivalent to second threshold limit (0·5 on 0–5 scale)
tR = minimum exposure time to achieve SR
ti = actual exposure time
c = actual concentration during exposure
K_2 = constant for receptor species
P_2 = constant for environmental conditions

Example Alfalfa exposed to 0·5 ppm SO_2 (c) for 20 hr (ti) where tR = 6 hr at 0·5 ppm SO_2 and $7 \cdot 3/K_2 P_2$ = 4·5 (under conditions of high humidity and saturation of soil with water);

$S = 0 \cdot 5 + 4 \cdot 5 \sqrt{0 \cdot 5 \, \log_{10} 20/6}$ = 2·1 (approx. 20 per cent foliar injury)

To allow for several fluctuations about the threshold limit more complicated functions are required to forecast injury.

Interactions between pollutants

Potentially, pollutant interactions are of great importance in causing damage. Such interactions may be expressed as:

$$\frac{A_1}{T_1} + \frac{A_2}{T_2} + \cdots + \frac{A_n}{T_n} = x$$

where A is the observed concentration or dose of each pollutant in a damaging mixture of pollutants and T is the known threshold for the same level of damage for each pollutant.[42] Thus if the combined action of the pollutants is greater than the sum of their independent effects (i.e. x is less than 1·0), then synergism is considered to occur. Similarly if x is equal to 1·0 or greater than 1·0 then the mixture is considered to be additive or neutralising respectively in its effects upon the receptor. Examples of all three interactions are given in table 9. Additivity is much the most common phenomenon. Claims of synergism are often found to be incorrect when judged according to the above criteria. For instance, figure 14 illustrates the combined effects of increasing smoking rates and air pollution concentrations upon human mortality due to bronchitis and to lung cancer. Both have been claimed to be the result of synergism between smoking and pollution.[146,155] However, only lung cancer is apparently a synergistic response, while bronchitis is apparently the result of an additive effect.

TABLE 9   Examples of pollutant interactions

Pollutants	Response	Interaction
Sulphur dioxide and smoke v. cigarette smoking	Human lung cancer	Synergistic
Cigarette smoking v. radiation	Human lung cancer	Synergistic
pH v. ammonia	Acute toxicity to fish	Synergistic
Sulphur dioxide and smoke v. cigarette smoking	Human bronchitis	Additive
Most water pollutants	Acute toxicity to fish	Additive
Sulphur dioxide v. hydrogen fluoride	Injury to plant foliage	Additive
Metal mixtures with zinc predominant	Acute toxicity to fish	Neutralising*
Fluorine v. calcium	Fluorosis in cattle	Neutralising
Ammonia v. sulphur dioxide	Injury to plant foliage	Neutralising

*Neutralising is only a partial effect, i.e. the mixture is still toxic

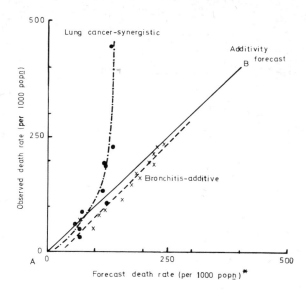

FIGURE 14   Interactions between smoking and air pollution in humans.[45]
*Forecasts assume additivity; thus the observed death rate for recep-
tors smoking twenty cigarettes per day and exposed to 250μg smoke is
compared with the forecast death rate, which is obtained by adding the
rates for non-smokers exposed to 250μg smoke and twenty-
cigarettes-per-day smokers exposed to no pollution. Any significant
deviation from the line of forecast additivity (A to B) indicates either
synergism or neutralisation

Additivity is a good working hypothesis when considering any mixture of pollutants.[42, 118] Bridges[156] has even suggested that it might be
applied to the assessment of all environmental genetic hazards,
including radiation, air pollutants and food additives. The main
objection to such an approach is that it presumes that all pollutants
operate in the same way and must be given equal weight as damaging
agents. Although this might be the case for simple mixtures[21,42] it
cannot be guaranteed in complex mixtures inducing subtle damage
over long periods of time.

Interactions between a pollutant and the environment can also
change the nature of a pollutant. For instance, inorganic mercury may
be transformed to much more toxic organomercury compounds in
bottom sediments of rivers.[157,158] Thus if the level of pollution is
measured by the inorganic mercury content of the water and sediments it is possible to underestimate pollution hazards.

Variations in environmental conditions
Many examples can be found in which the effects of air pollution, water pollution, pesticides[87] and radiation on plants and animals are modified by environmental stress factors. For instance, corrosion of metals[158] is determined mainly by the type of metal, sulphur dioxide concentrations, atmospheric salt content, duration of periods of surface wetness, temperature and wind speed, although at least five further variables may be involved. The corrosion rates of a given metal can be changed drastically by variations in wind speed, surface moisture, etc, without any change in the concentration of sulphur dioxide.

The best examples of definition of these variables can be found in fish toxicity studies, where the LC_{50} values for one or more pollutants in relation to temperature, pH, water hardness, dissolved oxygen, content, etc, can now be predicted with reasonable accuracy.

Variations in receptor condition
The sensitivity of a receptor varies in relation to (*a*) the health or condition of the receptor and (*b*) its stage of development or maturity. The former is often closely associated with environmental stress (e.g. water availability influences plant turgidity, which affects responses to sulphur dioxide). Generally, unhealthy receptors suffer more damage from pollution than their healthy counterparts.

One example of the influence of receptor development upon both sensitivity to and injury caused by air pollution has already been given (table 1; chapter 1). Other examples include the greater sensitivity of juvenile stages of fish to water pollutants, and the enhanced risk of radiation damage in pregnant females. The most sensitive sub-population of receptors (e.g. the young, pregnant females, adults in ill health) should ideally determine the minimum standards for pollution control.

Selective damage to particular stages of receptor development is also important in population survival. This problem is encountered in fishery protection, where control of damage to the whole life cycle ensures a viable fishery, although fishing is still feasible in some polluted waters receiving adults from adjacent unpolluted areas.[118,131] Constant recruitment thus gives an illusion of a viable fishery.

Responses in communities, etc
With one exception (table 6), all preceeding discussions relate to

individuals or small populations of receptors. The effects of pollution upon communities and ecosystems are much more complex, involving both direct and indirect damage. Such responses can be detected only by extensive experimental and survey research, especially when considering subtle responses to chronic pollution. The expression of total impact is best resolved as damage rates (e.g. x individuals/1000 population) for receptor populations or reductions in the productivity or diversity of the community (chapter 4). Such data are scarce and there are severe technical problems to be overcome in producing measurements of total impact that allow for anomalies in receptor distribution and the effects of pollution under all exposure conditions.

Nevertheless it has been demonstrated that declines in populations are logarithmically (or exponentially) and inversely related to dose.[160] Within a community, damage usually reflects the individual sensitivities of the component receptors,[161] a pattern maintained where doses exceed threshold limits for all receptors. Complications arise, however, where one or more of the following phenomena occur:

1   External factors (e.g. shelter, the immigration of less sensitive receptors), distort dose/response relationships. For instance, tree canopies can protect sensitive receptors by screening out air pollutants.[33]

2   Doses do not exceed the threshold limits for all receptors, thus permitting less sensitive receptors to occupy vacated niches within the community (e.g. limestone dust pollution encourages certain species of plants[162]).

3   Damage is the result of indirect or direct and indirect exposures to pollution which present no clear relationship between response and dose. For instance, pesticide pollution can result in damage to receptors *via* the concentration of residues as they pass up food chains. Also receptors can be damaged by a total pollutant dose composed of direct inputs (e.g. inhaled lead) and indirect inputs (e.g. the ingestion of fall-out lead deposited on vegetation).

4   Damage arises indirectly from the effects of pollution upon lower trophic orders, thus depriving an organism of its food.

The investigator of damage is faced, therefore, with the task of elucidating a series of complex relationships between environmental concentrations of pollutants, total dose, threshold limits and dose/response functions for each type of receptor with its various direct and indirect responses, and total damage. Formal expression of these relationships is impossible at present, owing to data and methodological deficiencies, but it will be essential for the accurate estimation of damage to communities and ecosystems.[10,133]

A few attempts have been made in pollution research to resolve these problems. For instance, some workers have studied the effect of acute exposures of plants to airborne dusts where damage is a mixed response to direct foliar injury and indirect soil modifications. Thus, Hajuk[160] proposed that

$$VT = m \cdot t$$

where VT = degree of direct and indirect effects of dust upon vegetation, m = grammes of magnesite dust per square metre and t = time in years. This was expanded to

$$SV = f(m \cdot t)$$

where SV = the structure of the community (diversity/productivity index) and f is a constant or complex function of climate, receptor sensitivity, soil conditions, etc. These and other formulae are, unfortunately, very specific to the receptors and pollution situation under investigation. The external variables (in this case, f) prove extremely difficult to specify throughout their ranges in relation to the dose/response function.

The prediction of doses from radioactive contamination has been formalised in relation to both marine and air pollution. In the latter case[77] a single or continuous emission of radioactive waste is related to total animal intake (I),

$$I = RfrE \text{ mg/day}$$

where R = pollutant deposited on vegetation per day (mg/km²/day), f = fraction of deposit remaining on vegetation, r = effective mean life (radioactive half-life) of the deposit, and E = vegetation grazed or eaten daily by the receptor or an intermediary in km²/day. The last factor can be converted to weight of vegetation and of pollutant (dose) consumed per day, given the ratio between area and crop production or weight of pollutant.

The contaminated vegetation may pass through an intermediary. Thus vegetation contaminated with 131-iodine is consumed by cattle whose milk becomes contaminated and is drunk by man. The actual dose delivered to man depends on the absorption/retention characteristics of the cow as well as on man's total intake of milk, but

$RfrEa$ mg/l = pollutant concentration in milk; and

$RfrEa$ mg/day = pollutant in milk per day,

where a mg/l per day is the amount of daily intake of 131-iodine appearing in each litre of milk. The ratio of atmospheric deposition and milk content can therefore be expressed as:

$$I : RfrEa \text{ mg/day}$$

These formulae require considerable modifications for different forms of pollution. For instance, where soil is contaminated the daily uptake by plants and the pollutant concentration in the edible parts must be computed. These may be functions of the deposition rate from the atmosphere or of seepage in drainage waters. In principle, similar formulae could be evolved for many pollutants that pass indirectly to receptors, given sufficient details of critical variables such as the proportion of surface contamination absorbed by the receptor.

The damage incurred by the receptor depends on whether or not the dose exceeds its threshold limit. In the case of elevated levels of 131-iodine this may be expressed as an increase over the spontaneous response rate (e.g. for thyroid cancer).

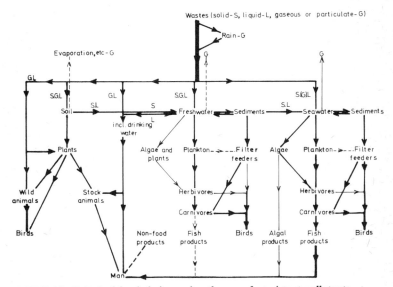

FIGURE 15   Principal food chains and pathways of persistent pollutants

For many persistent pollutants the data required to estimate doses
and subsequent damage are considerable, because more than one
route is involved in delivering the dose to the receptor. Figure 15
illustrates various possible sources and routes of such pollutants.[127]
Lead, for instance, enters man by inhalation of industrial and vehicle
emissions, ingestion of food grown in contaminated soil and ingestion
of polluted waters. For each route it is necessary to determine loadings, absorption rates at each trophic level and, in man, the doses
delivered to sensitive organs (e.g. the brain). Given these data,
estimation is still hampered by lack of dose/response functions, especially for subtle responses to chronic pollution among receptor populations (e.g. wildlife) of uncertain distribution.

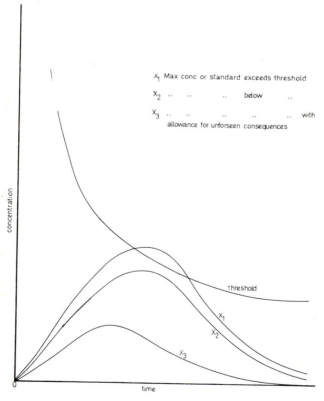

FIGURE 16   Environmental quality standards in relation to threshold or tolerance
            limits

SAFETY LIMITS AND QUALITY CRITERIA

The threshold limit offers an empirical base for environmental safety and quality criteria. Ideally the maximum permitted level of pollution in any given situation should be the first threshold limit (i.e. the maximum dose with no effect) of the most sensitive receptors under conditions of maximum sensitivity (figure 16). An additional safety margin might be added to allow for unusual habits, accidents and persistent pollutants with long-term effects. For instance, a value equivalent to 0·1 to 0·01 FTL would be satisfactory in most cases.

In practice, however, there are difficulties in setting environmental quality criteria quite apart from the general lack of threshold limit and dose/response data. Firstly, the selection of the most sensitive receptors involves subjective value judgements. Humans usually take precedence, although other receptors (e.g. wildlife in relation to pesticides) have been deemed acceptable. Secondly, what level of damage is to be regarded as acceptable? For instance, the hazards of the non-military uses of radioactivity are minute compared with current road driving hazards.[77,127] Yet to many people both hazards may appear to be equally important. Thirdly, what response is to be selected as a basis for a particular standard? Residual levels (e.g. pesticides in tissues) are infinitely preferable to lethal responses because they give some warning of damage. However, exposure to many pollutants is not associated with residual accumulations. Finally, there are technical, social and economic factors relating to the cost and operation of control measures which are often major determinants in the setting of environmental safety and control criteria. In fact many pollution control criteria do not exclude the possibility of damage, particularly as a result of long-term exposures to low-level pollution.

Most arguments about the appropriateness of a given criterion arise from technical uncertainties. For instance, various assumptions about the form of a dose/response function can produce quite different benefits from a postulated, incremental decrease in pollution (figure 17). The size of the expected benefit is a critical factor in the justification of stricter pollution control. Thus reliable threshold limits and dose/response functions are important in the setting of criteria and the evaluation of damage for pollution control purposes.

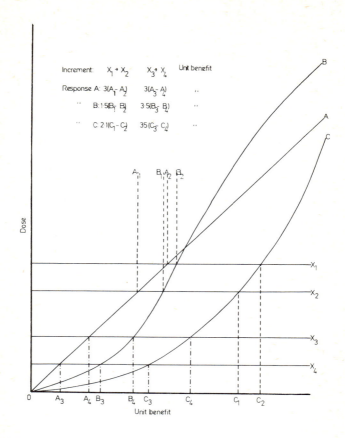

FIGURE 17 Dose/response functions and incremental benefits of reductions in pollution

CHAPTER 6

The estimation of damage: strategy and analysis

This chapter is an examination of the strategy of damage research and some approaches to the analysis of damage.[10]

STRATEGY

The purpose of damage studies is to measure the effects of pollution upon receptors, and may be separated into three main objectives:

1 The identification of responses to pollution.
2 The determination of the relationships between response and exposure to pollution (i.e. threshold limits and dose/response functions).
3 The measurement or estimation of the level of response (i.e. damage), preferably in terms suitable for social and economic evaluation.

To achieve these objectives attention must be given to:

1 The long-term effects of chronic pollution.
2 Interactions between pollutants and between pollutants and environmental factors which modify the effects of pollution.
3 Pathways of pollutants, especially persistent materials, through the environment and their contributions to the total pollutant doses received by receptors.
4 Monitoring of pollutants, receptors, environmental parameters and damage (chapter 4).

There are many ways of attempting to achieve the above objectives. However, the critical question is usually how to measure damage occurring in the 'field'. There are several approaches to this problem. These take the form of study types within which a variety of techniques (chapter 4) may be applied on almost any scale.

Scale of studies

For convenience pollution damage research can be segregated into micro- and macro-scale.

Micro-studies are defined here as including all research that has no spatial dimension. They are comprised mostly of experimental studies in the laboratory (e.g. toxicity tests, biochemical studies, accelerated corrosion tests for metals), which have contributed most of the quantitative information about damage (chapter 4), particularly in relation to threshold limits and dose/response functions. Such research is essential to understand the mechanisms by which pollutants affect receptors.

However, much of the data bears no obvious relationship to field conditions except where the conditions of exposure are similar and where attempts have been made to check the predictive accuracy of laboratory methods (e.g. fish toxicity trials[118]) by field experimentation. Furthermore many data relate to acute pollution. Greater emphasis should be placed upon studies of sub-lethal damage due to chronic pollution and the development of laboratory tests for predicting damage.

Macro-studies enbrace all field experimental and survey techniques that involve some spatial relationship between pollution and damage. Basically they fall into four main categories:

1 *Area studies*[42] in which the effects of one or more pollutants upon receptors are studied within a pre-selected area. They may range in dimension from global to very localised (e.g. a pond) studies. Generally, the accuracy of damage measurement increases with decreasing size because the variation in exposure conditions, etc, decreases and the logistics of the study improve. For larger-scale investigations the most suitable areas are those with reasonably well defined topographical boundaries (e.g. catchments, airsheds).[37] For more detailed biological research, however, smaller study areas are required (e.g. lakes, small woods). At this level emphasis needs to be given more to study sites that contribute to the wider understanding of pollution damage, rather than to sites of unique importance (e.g. for conservation reasons).

2 *Single-source studies*[163] in which the effects of one or more pollutants from a particular activity (e.g. steel making) are investigated. The spatial limits are defined by the range of the pollutant from the source. By concentrating upon isolated sources it is possible to reduce the number of other factors (e.g. other pollutants) affecting damage. This approach permits research in depth, particularly if the co-operation of industry, controlling authorities and the public is obtainable.

3 *Single-pollutant studies*[164] in which the effects of a pollutant upon one or more receptors are investigated. Practical field studies may be limited in their spatial dimensions, even to single sources, thus falling into the previous category.

4 *Single-receptor studies* in which the effects of pollution upon a specific receptor are studied in depth. The objective of this approach is the assessment of hazards to a particular receptor. Its conservation and possible use as an indicator of pollution (e.g. lichens and sulphur dioxide[165]) may also be important considerations. Such studies can be conducted within confined areas, around single sources and in relation to single pollutants.

All four approaches produce valuable damage data. They also serve as pilot or case studies for damage appraisal on a wider scale and help to develop a national picture of damage in similar areas.[19] Single-source studies are especially useful in studying the environmental impact and control problems of an industry or a sector of the economy.[74] An important point to remember about all these approaches, however, is that most pollution is chronic in nature, with subtle long-term effects. To measure the range and severity of damage caused by such pollution requires long-term, integrated experimental and survey research which is expensive and will not immediately provide unequivocal results.

Desk studies

To establish research and control priorities it is helpful to review a pollution problem by the desk study method, in which little or no practical research is conducted. Most of these reviews[62] are short-term versions of the case studies discussed previously. They are designed to elucidate the determinants of pollution, its impact and control problems in relation to specific industries,[74] waste disposal,[166] pollutants,[167,168] media[55,44] or geographical areas.[37,63] The principal characteristic of such studies is their reliance upon existing data on wastes, pollutants, receptors damage, etc, the deficiencies of which (chapters 2–5) severely limit the estimation of damage. However, the determination of these limiting factors is vital in the evaluation of research priorities.

Data sources   Case studies rely upon the available sources of data relating to waste production, pollution levels, receptor distribution, environmental conditions, land use, damage rates, etc. Table 10 lists relevant data sources at their various administrative levels.[37,169] Much unpublished information about pollution and damage is held by local and controlling authorities (e.g. parks and gardens departments, the public health inspectorate), and by local industries. This can be obtained most easily by personal interviews and the use of postal questionnaires.[37]

TABLE 10 Sources of waste, receptor, pollutant and damage data in spatial studies of pollution damage [e.g.37]. Examples of sources providing direct data on wastes, pollution and damage, and also indirect data suitable for computing potentially polluting loads.

Specific source: may cross authority boundaries, e.g. airports, industrial works, industrial trade associations.

Specific interest: may cross authority boundaries, e.g. conservation groups, angling clubs, recreation organisations, university research groups.

Minor local authority area: districts (formerly cities, boroughs, rural and urban districts), e.g. public health service, parks and gardens departments, Medical Officers of Health (may cover combined areas), agricultural data,* public analyst.

Major local authority area: county or metropolitan area, e.g. planning authority, county MOH, public analyst,* waste disposal officers.

Other authorities, government departments and agencies at regional or sub-regional level: e.g. Agricultural Development and Advisory Service,* public transport authorities,† Forestry Commission, Nature Conservancy Council, planning councils.

National government departments: e.g. Department of Health and Social Security, Office of the Registrar General†

Government or government-funded organisations with specific local projects or interests: e.g. Department of the Environment (road research, special projects), National Survey of Smoke and SO_2,* Alkali Inspectorate, Research Councils, Nature Conservancy Council.

Organisations orientated by topography or function: e.g. Regional Water Authorities,† Sea Fisheries Committees,† Central Electricity Generating Board,† Atomic Energy Authority,† National Coal Board,† Alkali Inspectorate, Radiochemical Inspectorate.

*May collect data on a smaller or local authority basis.
†Regional, but organised either on district or parish basis or by source group.

Use of data Pollutant and receptor data are used to prepare maps[37,169] of their distribution about a specific source of pollution or within a pre-selected area. The objective is to determine the numbers of receptors exposed to various doses of pollution. Using known threshold limits (chapter 5), it is then possible to indicate zones where damage may be anticipated. Similarly, using known dose/response functions it may be possible to estimate the rate of damage (e.g. x dead/1,000 population) and the quantity of receptors damaged (e.g. y tons' yield lost per annum). Because of data limitations, however, it is rarely possible to map pollutant distributions accurately.[37,169] Furthermore reliable threshold limits and dose/response functions have been determined for very few responses and receptors. Any estimate made should then be checked against existing data of damage within the study area.

Tables 11 and 12 and figure 18 give crude examples of estimated

TABLE 11 Estimated yield losses in grasslands due to sulphur dioxide pollution in Greater Manchester[37,44]

Receptor group	Losses per annum			
	Yield %	Acreage	Tons Yield	Estimated value (£000)
All grasslands (agricultural, permanent, recreational, cemeteries, etc) – 100 per cent sensitive	8·6	8,100	12,500	500
All grasslands – 30 per cent sensitive	2·6	2,480	3,730	170
All agricultural grasslands – 100 per cent sensitive	7·7	6,400	10,600	430
All agricultural grasslands – 30 per cent sensitive	2·3	1,920	3,180	130
Agricultural seeded pastures – 100 per cent sensitive	7·4	6,230	10,230	385
Agricultural seeded pastures – 30 per cent sensitive	2·2	1,870	3,070	115

Criteria (a) Visible injury at 200–400 μgSO$_2$/m^3 air (summer peak daily concentration) causes 1 per cent yield loss and $>$ 400 μg SO$_2$/m^3 air causes 5 per cent yield loss.

(b) Cryptic injury at 100–200 μg SO$_2$/m^3 air (summer average daily concentration) causes 10 per cent yield loss and $>$ 200 μg SO$_2$/m^3 air causes 40 per cent yield loss.

Value of losses These are extremely crude estimates of direct costs of damage. Assuming 30 per cent sensitivity, total losses for all agricultural grasslands are estimated at £145,000–£200,000/annum, for all agricultural and horticultural crops approximately £400,000/annum and for all cultured plants (including those in parklands, etc) approximately £500,000–£700,000/annum. The last estimate is very approximate, but with indirect costs it could exceed £1 million/annum.

TABLE 12   Calculation of the possible contribution of a high-level emission in a multi-source area to local mortality (bronchitis) [45, 169]

(a)   *Data*

Pollutants	$\mu g\ m^3/air/24\ hr.$		Duration (%)
	Smoke	*SO₂*	
Average (winter)	100	140	94
Peak (winter)	360	520	6

Estimated contribution (ex source)

Average (winter)	10 (10%)	7 (5%)
Peak (winter)	72 (20%)	52 (10%)

Bronchitic mortality (males): annual rate per 1,000 population per annum

Comparable rural rate	4
Local rate	9
Local excess	5

(b) *Calculations*

Threshold: 250 μg smoke and 500 μg SO₂ (lower values may be used)	
Contribution *ex* source to above-threshold levels (%)	15%
Allowance for cumulative effect of below-threshold exposures	5%
Total contribution	20%

Assuming local excess of mortality is due solely to pollution, then the contribution of this source

$$= 5 \times \frac{20}{100} = 1\ \text{death}/1{,}000/\text{yr}$$

These estimates are extremely crude and are based on simplified dose/response relationships with no allowance for other pollutants, occupational exposures, habits,* socio-economic status, etc.

For example – allowing 75 per cent for contribution of smoking, the estimated rate would be 0·25 per 1,000 population.

damage.[37, 169] As table 12 indicates, it is necessary when dealing with single sources in a multi-source area to calculate the contribution of the individual source to total pollutant concentrations and to total damage. Simple dilution and diffusion formulae[170] may be used to estimate such contributions, but the process of diffusion is often so complex that the resulting estimate is of very dubious validity.

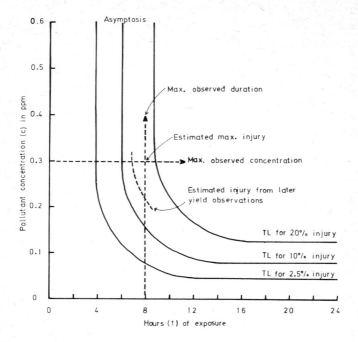

FIGURE 18   Example of the estimation of yield loss due to episodic air pollution from a single source.[45,169] *Situation.* Pollutant emitted for five hours. Maximum observed duration of pollutant at peak concentration of 0·30 ppm for eight hours at one mile downwind of source during some previous episodes. From known threshold limits this corresponds to a maximum injury to crop growing one mile downwind of 16 per cent foliar destruction. Assuming the percentage of acute injury is roughly equivalent to percentage loss in yield (chapter 5), then the estimated yield loss is between 12 and 18 per cent compared with an observed loss of 13 per cent

In the few cases where damage data (e.g. human mortality[17]) are available attempts may be made to explain local variation in damage rates in terms of pollutant doses.[17,37] Inevitably these are complex statistical exercises, because it is necessary to account for many other variables (e.g. population density, socio-economic status) which may be related not only to damage but also to pollution levels, and to each other.[37] The determination of the proportion of a response due solely to pollution is therefore extremely difficult, and it is not surprising that there is considerable uncertainty in estimates of pollution damage costs (tables 11 and 13). Additional complications arise from the cumulative effect of chronic pollution. For instance, damage rates may be more closely related to past than to current concentrations of pollution.[37] This point is rarely investigated in short- or long-term case studies of damage because it requires data from previous years, at least for pollutants. These data are essential for establishing recent trends in pollution and damage (chapter 2). Trends can be also used to forecast future levels of pollution and damage. This is a complex exercise[10] however, especially for damage, mainly because it is difficult to forecast future distributions of receptors. Often it is safest to assume that receptor populations will not change significantly, at least in the immediate future.

TABLE 13   Estimated current total damage costs due to air pollution in the UK[10, 19]

| Item | Costs (£million/annum) | | | |
| | Economic | | Social | |
	Mean	S.D.	Mean	S.D.
Painting	–	–	6·3	2·5
Laundry, dry cleaning, etc	0·5	0·1	164	60
Exterior cleaning of buildings	–	–	1·5	0·2
Window cleaning, etc	5·0	0·2		
Corrosion and protection of metals	42	28	–	–
Damage to textiles and paper	33	4.5	–	–
Agricultural production	195	110	–	–
Amenity—damage from:				
Point sources	–	–	100	33
Motor vehicles	–	–	3	1
Health	130	63	510	243
Rounded totals	410	130	(780)	(252)
Grand total		1,190		

Figures rounded to nearest whole number. (   ) = incomplete estimates.

The high levels of standard deviation (SD) clearly indicate the uncertainty and lack of reliable data of physical estimates of damage.

Where no data are available for pollutant concentrations, etc. it is possible to use wastes (e.g. animal faecal wastes, vehicle emissions, mining spoil) as proxies for pollution. By establishing their distribution, hazards to amenity (e.g. odours, visual intrusions), health (e.g. vehicle emissions of gaseous pollutants) or water quality[37,45,55] may be implied. These proxies cannot, however, be directly related to damage and can be regarded only as indicators of pollution.[55]

Complaints about pollution problems by the public to local and controlling authorities and to industy are useful indicators of pollution and damage. But they have certain limitations.[37] Firstly, unless they arise in response to overt damage they are subjective reactions to nuisances. Secondly, rates of complaint by the public are influenced by local factors such as socio-economic status, awareness of pollution, acclimatisation to pollution, and the employment of some human receptors at the source. Thirdly, complaints tend to relate strongly to obvious and specific sources of pollution. They are most useful in identifying the range and frequency of an intermittent pollutant (e.g. noise) from a specific source and its impact upon local, residential amenities.

Value of desk studies   Desk studies are limited in their accuracy by the deficiencies of existing data and monitoring systems. Nevertheless they do indicate gaps in information and priorities for future monitoring, damage research and pollution control. Also, they may provide sufficient data and information about fundamental relationships (e.g. coefficients for waste production or pollutant levels per head of population) to serve as bases for empirical modelling studies[10,37]

ANALYSIS
The types of study discussed in the previous section have similar data requirements and analytical procedures, despite their differences in approach and subject matter. By drawing together these common points it is possible to devise a general analytical framework for damage appraisal.

An analytical framework
Figure 19 outlines an analytical framework for the appraisal of pollution damage. Ideally the monitoring and other information systems would provide accurate data for each stage. Thus it would be possible to determine the characteristics (e.g. chemistry) and output of the

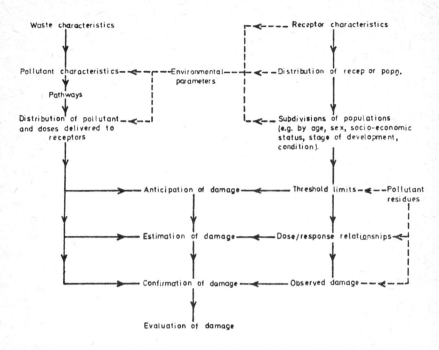

FIGURE 19   An analytical framework for the estimation and appraisal of damage

waste, its interactions with and pathways through the environment, and their relationships with the resulting levels of pollution. The system also permits the determination of contributions from each source to the total pollutant dose. At the same time the characteristics (e.g. age structure, condition) of the receptor and its distribution would be determined, together with their interactions with environmental factors, especially those affecting sensitivity to the pollutant (chapters 4–6).

The main objective of the exercise is to determine the pollutant dose delivered to the receptor. The distributions of dose and receptor populations are therefore compared. Ideally, accurate threshold limits and dose/response functions (chapter 5) would be available from experimental exposure and field studies. Zones of anticipated damage within the study area would therefore be delineated where doses exceed relevant limits.[37,45] The severity (e.g. percentage yield loss) or rate (e.g. deaths/1,000 population) of damage could be estimated within the same zones by reference to known dose/response functions. The estimation of absolute damage (e.g. total deaths) is a simple multiplication of the number or quantity of receptors times the damage rate within each zone. Zone estimates are summed to derive a total value for the whole study area (table 11).[37]

Zonal estimates serve as hypotheses to be tested against observed damage levels. Reliable data of actual damage are uncommon; most short-term case studies cannot progress beyond this stage. In longer-term studies, however, survey techniques may be used to determine actual damage levels within the study area.

The pitfalls of survey techniques have been discussed already (chapter 4). However, care must be taken to ensure that: (a) the exposed sample is representative of the receptor population and exposure conditions; (b) the control (unexposed) populations are as similar to exposed populations as possible; (c) the investigators are competent in the recognition and measurement of symptoms; and (d) the survey covers critical variations in pollutant dose, environmental conditions (e.g. seasonal changes and receptor development).

The use of time-series data for at least the last ten years in respect of pollution, receptor distribution and damage is implicit in the framework (figure 19 – not shown). These data are used to determine trends in all three parameters and, by extrapolation, to forecast future damage. They are also essential in the interpretation of the relationship between pollutant dose and damage. For instance, past peaks of pollution may still be affecting current damage, the size or distribution of existing receptor populations and the concentrations of pollutant residues within receptors.[68] The analyses of survey data require comprehensive statistical analysis[37] to indicate both the relationships between pollution and damage in relation to time, and the influence of other environmental factors upon damage. Simple correlation tests are insufficient to handle such complex situations (figure 12).

The framework assumes ample support facilities for experimental studies to test the validity of field observations. Such experimental studies have a fivefold function: (a) to define the responses of the receptor to the pollutant, including the relationship between injury and damage (chapter 1); (b) to investigate the mechanisms by which the pollutant damages the receptor; (c) to determine or confirm threshold limits and dose/response functions under conditions comparable with those in the field; (d) to explain discrepancies between estimated and observed damage (e.g. due to pollutant interactions);

and (*e*) to determine the relationships between pollutant dose, residue concentrations in the receptor and damage.

Multiple pollutant/receptor situations   The framework is designed to analyse the interactions between a single pollutant and a single receptor or receptor population. In the external environment, however, several pollutants may occur in one area and may damage more than one type of receptor. In such cases preliminary screening may be attempted, using a matrix format (figure 20) which includes all possible pollutant/receptor combinations[171] – a technique especially useful in short-term studies. (These are widely used in pollution impact assessment studies in the USA.) The matrix permits the elimination of unlikely combinations and, if the remainder are ranked according to some arbitrary scale of suspected impact, the determination of priorities for detailed experimental and survey research.

Receptor	Pollutant	Suspended solids	Oxygen demand	Heat	Scrap and other solids	Ammonia	Zinc	etc.
Biota	Fish A	-	3	-	-	2	1	
	Humans	-	0	-	-	0	0	
	etc.							
Abstractions for water supplies	Industry CB	-	1	-	-	1	0	
	Town B	-	3	-	-	2B	3B	
	Agriculture B	-	0	-	-	0	0	
	etc.							
Amenity	Angling	-	3	-	1	2	1	
	Swimming	-	3	-	1	0	0	
	etc.							

FIGURE 20   A preliminary screening matrix for combinations of individual pollutants and receptors in a single medium.[169,171] Ranking may signify merely presence or absence, or some arbitrary scale of apparent impact. *A:* further segregation into species may be feasible. *C:* further segregation into individual industries may be feasible. *B:* if these abstractions operate, impact can be assessed directly. On the other hand their absence may be due to pollution necessitating indirect appraisal of impact in terms of utilisation of alternative sources

Preliminary screening may indicate interactions between pollutants (chapter 5). As a working hypothesis at the screening and damage estimation stages, additivity may be assumed unless synergism or neutralisation is indicated by work elsewhere.[42,172] Each group of interacting pollutants or their interaction product (e.g. photochemical smog) is treated as a single pollutant.

The analytical framework may be applied to most damage studies. It has two main advantages. Firstly, it identifies data requirements, and therefore data deficiencies, for damage estimation and measurement. Secondly, it offers a systematic procedure by which estimates of damage can be prepared and checked against measurements of observed damage.

Some major technical problems

Whatever broad approach to the study of pollution damage is used, many technical problems will be encountered. Most of these have been indicated in previous chapters. There are, however, four major areas of difficulty:

1 Estimation of damage due to episodic pollution.
2 Analysis of multi-route pollution and its effects.
3 Analysis of '*ex post*' situations – compensation and recovery.
4 Damage to complex receptors.

Episodic pollution The major problem with episodic (i.e. occasional or intermittent) pollution is the uncertainty of its incidence in time and space. Almost all pollution episodes result from accidents during the transport, storage and disposal of wastes and products, especially in the metal, chemical, oil and nuclear power industries. Severe production accidents are much less common. Studies of polluting accidents must be orientated towards specific pollutants or industries on a wide scale (e.g. global or national) in order to obtain sufficient data for determining the statistical probability of their occurrence.

Industrial operations imply a finite degree of risk of environmental damage inherent in mechanical failure (e.g. breakdowns, power shut-down), maloperation (e.g. poor supervision, sabotage) and natural disruption (e.g. flood, bacterial contamination).[78,117] The path taken from an event initiating an accident is one of many, each path and each phase with its own probability of occurrence. The assignment of such probabilities depends upon accurate knowledge of the performance of every item and operation in the industrial plant.

Analyses may even include estimates of delays and partial failures in sub-systems (e.g. cooling and safety systems), and their deterioration under stress.[77,78]

In some cases (e.g. oil tanker collisions and spills[117]) the frequency of actual accidents can be used to calculate the probability of future incidents. In others (e.g. nuclear reactor blow-downs) where experience is limited and accident probabilities are very low the latter can be estimated only from the probabilities of each failure sequence. The damage investigator is interested ultimately in the likely frequency and magnitude (viz. quantity of pollutant) of a polluting accident. These may be defined as probabilities of frequency for events of increasing magnitude. For each magnitude the range of or area affected by each event (table 14[117]) can then be estimated. This may be refined into zones of decreasing dose from the point of discharge.

TABLE 14   Some data from an analysis of discrete marine oil spills[117]

	Common characteristics of spills	*Some average or median values**
Source	75% of discrete spills from tankers	–
Composition	90% composed of crude or residual oils	–
Volume	70% > 5,000 barrels	Average 30,000 barrels
Distance offshore	80% < 10 miles	About one mile
Duration	75% longer > 5 days	Twenty days
Range	80% reach < 20 miles of coastline	Cover three to four miles of coastline
Type of coastline	85% spills off recreational shores	–
Distance of spill from port	75% < 25 miles to nearest port	About eight miles
Clearance time available before oil reaches coastline	80% spills occur < 1 day	Less than two hours

*Values heavily weighted by a few major incidents (e.g. the *Torrey Canyon*).

Further adjustments may be made to allow for variations in disposal (e.g. height of emission) and environmental (e.g. wind direction) conditions. Thus it is possible to determine the probable limits of distance and area affected and of doses delivered to receptors under a variety of environmental conditions.

Given the threshold limits and dose/response functions for each receptor, a casualty or damage rate for receptor populations within each dose zone can be estimated (e.g. induction of thyroid cancer by 131-iodine[77,78]). This may be extended as a series of probabilities of events of increasing casualty rate (i.e. X_1 to X_n per 1,000 of the population).

Studies in the nuclear power and in other industries (e.g. oil transport[117]) suggest that the principles of accident analysis have wider application in the study of episodic pollution from waste disposal, manufacturing, storage and transport operations. The spatial and temporal distributions of mobile units (e.g. road tankers) present severe analytical problems, however, unless they are grouped into zones of unit movements, each being considered as a stable point source (e.g. as a nuclear reactor) of possible pollution.

Multi-route pollutants Persistent pollutants (e.g. metals, some pesticides, fluorides, radioactive materials) travel several routes through the environment. Receptors may, therefore, receive doses of a pollutant in more than one form and from one or more sources.[127] Accurate information about such parameters is valuable in damage estimation and pollution control. Often, however, knowledge is confined to broad outlines (e.g. of pathways), interspersed with quantitative information about specific parameters (e.g. pesticide residues in birds[68]). Even then the relationships between various phenomena (e.g. pesticide residues and dose rates, residues and breeding failure rates) are far from clear (chapters 2, 5).

One useful technique in such circumstances, especially in specific source or pollutant studies, is critical pathway analysis. A classic example can be seen in the studies of radioactive wastes discharged from Windscale into the Irish Sea.[75,77,173,174] The objectives of critical pathway analysis are to identify:

1 The routes of pollutants through the environment from the point of discharge to receptors. These may involve both physico-chemical and biological diffusion processes. Particular attention is paid to those pathways along which pollutants are concentrated by physical (e.g. sedimentation in estuarine muds) or biological

(e.g. concentration by seaweeds) processes. Primary, secondary, etc, pathways may be distinguished, depending upon the hazard or dose presented by each pollutant to the receptor.

2   Critical materials—those materials which are contaminated by and concentrate the pollutant, thus offering the greatest hazard to the receptor (e.g. man by contact or by ingestion).

3   Critical population or group—those receptors exposed to the pollutant at the end of the pathway. These may be subdivided into groups of varying sensitivity, dose rate, etc.

4   Critical organ—the organ or tissue of the receptor most exposed to and likely to be damaged by the pollutant.

The technique thus seeks to identify the relationship between the quantity of waste discharged and the maximum dose and risk to which receptors are exposed.

Critical pathway analysis involves intensive monitoring initially to establish the various routes, etc, and continuous monitoring of key points to check the stability of the diffusion patterns. However, it is invaluable for close control of specific discharges of persistent pollutants. The discharge rates have known relationships with dose rates or exposure levels. It is possible, therefore, to calculate maximum permitted discharge rates that do not exceed accepted exposure limits[29] for critical populations. Critical pathway analysis also facilitates the identification of biological indicators of pollution at key points along the pathways.[75, 173, 174]

The principles of critical pathway analysis have wider applications to the study of diffusion and damage by non-radioactive persistent pollutants (e.g. metals, organic chemicals) derived from specific industrial waste disposals. Studies of persistent pollutants from more diffuse sources (e.g. lead from vehicle emissions) are feasible in so far as they might identify major pathways and their contributions to total doses delivered to receptors (figure 15).

'Ex post' situations   Pollution damage is usually assessed by direct comparison of exposed receptors or communities with those in similar but unpolluted areas. The quantifiable difference (e.g. in reduced growth or numbers of receptors) is frequently considered a suitable measure of damage (chapter 4). However, the majority of acute and chronic pollution problems are not new situations, i.e. the investigator is faced with an historical or *ex post* situation. In such situations, especially where chronic damage occurs, there may be subtle changes amongst receptors and communities that can result in over-

or underestimates of total damage. For instance, within a community the elimination of sensitive receptors by pollution may be partially or wholly compensated for by the growth of less sensitive receptors which colonise the vacant ecological niches.[132] This compensatory development can differ in quality and quantity (e.g. yield of grass species) from the product of the original receptor mix. Over longer periods of time sensitive receptors may evolve less sensitive strains, with similar results. In terms of total damage, therefore, the orthodox comparative approach *underestimates* damage to the *original* receptor community by concentrating on current yield or quality losses. To cope with such situations an index of change in community and ecosystem structure (e.g. diversity of species) is required.[10, 133] More complex situations can arise. For example, it is probable that the sensitivity of plants to chronic sulphur dioxide pollution is greatly affected by the sulphur nutrition of the receptor.[175] Long-term pollution might thus result in improved yields where soil sulphur is deficient[32] and depressed yields[4] where soil sulphur is sufficient or in excess of the receptor's requirements.

These phenomena suggest that normal survey techniques are inadequate for measuring the true extent of damage. Much more intensive ecological studies are required, together with laboratory and field experimentation, to evaluate the nature and rate of subtle changes. Similar approaches are needed to two other phenomena encountered in *ex post* situations.

Firstly, it is often impossible to distinguish between the initiation and the exacerbation of a disorder by pollution unless prospective techniques are employed (chapter 4). For instance, although it is known that young children suffer bronchial disorders due to air pollution,[176] which is also a major contributor to mortality and morbidity due to chronic bronchitis in middle and early old age,[16, 17] the relationships between the two responses are obscure. Do the children suffering bronchitis form the bulk or the core of the adult patients? If so, is this a cumulative effect of pollution and other environmental factors alone or does it have some genetic connotations?

Secondly, acute pollution has declined in many areas; recovery from such pollution presents some interesting examples of ecological recolonisation and succession by sensitive receptors previously excluded by higher levels of pollution. In such circumstances two questions may be asked: (*a*) what is the quantitative rate of recovery of receptors and communities in relation to decreases in pollution; and (*b*) will recovery ever be complete? Data relating to both issues are limited. There are examples of recovery among individual receptors (chapter 5) which suggest that the process is a non-linear function of the time since pollution abated. As a working hypothesis the recovery period might be assumed equivalent to the exposure period required to cause damage at the original concentration of pollution.[45] However, surviving receptors may exhibit damage (e.g. advanced mortality[129]) after the complete abatement of pollution, with possible long-term repercussions upon later generations.[68,88]

Complex receptors   Frequent references have been made to the difficulties of measuring damage to complex receptors (i.e. communities, structures and ecosystems). Relevant threshold limit and dose/response data are lacking. In the case of natural communities and ecosystems there is also a gross lack of information about their structures and dynamics. Much ecological research is required to remedy this deficiency.

Unnatural systems present somewhat different problems. Structures (e.g. buildings) are composed of several receptors (e.g. brick, lead, steel, paint), all with different responses and thresholds to each pollutant. Most of the latter can be measured according to technical criteria (e.g. depth of corrosion, loss of strength). The effects of pollution upon a whole structure, however, influence its useful life and maintenance demands.[10,171] Damage is therefore measured directly in economic terms relating to decreases in the value of the structure and increases in maintenance costs (table 15).[171]

The technicalities of the economic evaluation of damage are not the subject of this text.[9,10] However, changes in property values (which are sometimes used as damage indicators) are complicated phenomena which reflect not only direct damage by pollution but also a large number of environmental and other factors. Additional problems arise with materials such as fabrics, where fashion may dictate early replacement before the physical damage becomes apparent.

The evaluation of maintenance costs is complicated by routine maintenance operations. The costs of cleaning, in such circumstances, may not vary with the level of pollution and so 'mask' the damage caused. These and other phenomena (e.g. the influence of socio-economic status upon house and clothing maintenance costs) deserve further investigation to improve the accuracy of measurements of pollution damage to property.

Pollution damages facilities for recreational and other amenities

TABLE 15   Calculation of corrosion maintenance costs for elevaι
storage tanks[171]

First estimate of tankage based on volume data

1 Total estimated storage in elevated steel tanks is $11,000 \times 10^6$ gallons.
2 Typically elevated storage tank = 1×10^6 gallons (1 Mg).
3 Equivalent number of 1 Mg tanks = 11,000.

Second estimate based on steel plate tonnage

4 Total tonnage of steel plate in all municipal water systems is 7,100,000. Of this, some 70 per cent is estimated to be in elevated storage tanks.
5 A 1 Mg tank is about 350 tons.
6 Number of 1 Mg tanks in service is $0.70 \times 7,000,000 \div 350 = 14,300$.
7 Average of items No. 3 and No. 6 = 12,700.

Calculation of pollution cost

8 Area of hemisphere + roof = $10,050 + 5,020 = 15,070$ ft². Other designs range from 12,000 to 30,000 ft². Use 20,000 ft² for calculation.
Annual extra maintenance cost = $0.0167 ft²/yr at typical urban corrosion rates for typical steel plate in use.
$20,000 \times 0.0167 = \$334$/yr for 1 Mg tank.
9 Based on observation, about 80 per cent of the elevated tanks are exposed to contaminated atmospheres.
Total annual extra cost caused by pollution
= $0.80 \times 12,700 \times 334$
= \$34,000,000.

(e.g. pollution of coarse fisheries). The lack of a common system for classifying the diversity of facilities and their uses is a fundamental obstacle to research. Systems based upon ecological, physical, ownership, usage, spatial and economic criteria have been suggested.[177,178] None of these is acceptable for most environmental research purposes. Furthermore facilities compete *with each other* and between resources *within* themselves. These interactions have not been analysed to any significant extent. Also there are methodological objections[9,10] to existing techniques[179] of evaluating recreational demand which are based upon the visit rates and distances travelled by each visitor in relation to the costs of travel. Finally, no satisfactory means of determining the social value of a facility and its depreciation due to the effects of pollution has yet been devised.

The preceding discussion refers only to four major difficulties in pollution damage research and appraisal. Previous chapters have raised many others of varying magnitude. All are symptomatic of a chronic lack of knowledge about pollution and its effects on the environment. This deficiency will not be met effectively by piecemeal research. A more positive and comprehensive approach is required. It would be a useful start to consider pollution as only one of many environmental parameters affecting receptors, communities and ecosystems. Whatever scale of study is envisaged, this implies a need for integrated programmes of experimental and survey research of the polluted environment.

CHAPTER 7

General conclusions

Earlier chapters reviewed the state of knowledge about pollution and its effects upon the environment. It was apparent that there are many gaps in existing information about pollution damage; much of the available data is of limited use, owing to various inaccuracies. To a certain extent these deficiencies are attributable to the limitations of various methods of measuring damage. There is also a lack of both clear objectives and a coherent strategy for damage studies. Subsequent chapters examined these issues and the problems which must be resolved to obtain the sound estimates of damage essential for rational policy decisions in environmental conservation and pollution control.[10, 11]

The objectives of damage studies can be summarised as the identification of the responses of receptors to pollutants, the determination of relevant threshold limits and dose/response functions throughout the full range of environmental conditions, and the measurement of damage caused by actual exposures to pollution in terms suitable for social and economic evaluation. The estimation of damage requires, however, accurate data of: (a) the levels and distributions of pollutants, including their proportions derived from various sources, together with any interaction products; (b) environmental factors (e.g. temperature, wind speed) influencing pollutant diffusion and receptor distribution and sensitivity; and (c) the distribution and condition of receptors, including communities and ecosystems. With these data, threshold limits and dose/response functions may be used to estimate damage in the field on a wide scale.

The strategy required to achieve such estimates of damage (chapter 6) may be resolved into three main approaches to pollution research.[10] The first concerns global pollution, where attention should be focused primarily upon the diffusion of pollutants and the interactions between their various effects upon world climate, life-support systems and ecology. To these ends it will be necessary to improve global monitoring systems of 'baseline' and polluted environments.[14, 180, 181] The second approach concerns the study of damage caused by episodic pollution arising from accidents. The primary objective should be the determination of the probable frequency, magnitude and distribution of accidental damage. Studies of individual industries would play a major role in such work, which will require considerable help from the mathematical and engineering disciplines. The third and most demanding type of research concerns the study of damage due to long-term chronic pollution. Single-pollutant and specific-source case studies permit detailed investigations of damage and assist greatly in pollution control (chapter 6; e.g. critical pathway analysis). Similarly, specific receptor studies may be used in the appraisal of conservation problems. However, the most flexible approach is the area study, which can be applied on any scale and can include any number of sources, pollutants and receptors.

Whatever approach is made, attention will have to be paid to baseline (unpolluted) as well as polluted situations in order to determine the true extent of damage.[14,88,181] Furthermore the long exposure times required to induce subtle forms of chronic damage imply research programmes of a long-term nature, including the use of prospective survey techniques. It is essential that experimental research should utilise where possible the same basic material and environmental conditions as are encountered in the field. This will ensure more representative results and greatly improve the predictive value of many toxicity testing methods. The relationship between long-term exposure and chronic damage, particularly in areas suffering long-standing pollution or where multi-route pollutants occur, requires much more careful attention to interpretation than it has hitherto received.

All three approaches suffer from a common limitation; it is easier to study pollution damage to individual receptors than to communities and ecosystems, because no satisfactory method of measuring damage to complex receptors has yet been devised.[133] Area and specific source studies of damage offer considerable scope for research into the same problem, but some fundamental ecological research and development of techniques are also required.

When examining priority topics for damage studies it is necessary to consider the relative significance of each pollutant in terms of trends in its environmental concentrations and the damage that it causes. Chapters 2 and 3 briefly reviewed available information on these matters. Owing to lack of data the determination of priorities is difficult, and there is no objective method of comparing the relative significance of pollutants affecting different receptors in the same or in different media. The brief summary which follows is a personal view of pollution research priorities.

1   In terms of episodic pollution, greatest attention should be given to the study of accident damage arising from the manufacture, storage and transport of oils and chemicals. The standard of accident analysis should be greatly improved.

2   The effects of persistent metals, industrial chemicals, radioactive wastes and persistent pesticides upon ecosystems and life-support systems are aspects of global pollution deserving priority attention. At a lower-level of priority, efforts should be made to determine in detail the dynamic effects of carbon dioxide, aerosols and particulates and possibly SST operations upon the global climate. (Much relevant work is already under way.)

3   Metals (especially lead, cadmium and mercury), persistent pesticides and industrial chemicals are the most important pollutants causing chronic contamination of all media, with potentially serious effects upon human health, wildlife and especially marine biota and ecosystems. Certain radioactive materials (e.g. tritium) might be included in this group. Generally, much more information is required about the diffusion (physico-chemical and biological) and pathways of such pollutants through the environment and their long-term subtle effects upon receptors. We lack much information on the toxic mechanisms of these pollutants in various receptors.

4   Eutrophication, especially of impounded waters, is important in terms of water quality, local amenities and the survival of biota (e.g. fisheries). The long-term effects of gaseous and particulate atmospheric polutants upon human health deserve research in depth. However, evidence is accumulating to suggest that plants are more sensitive to these pollutants than are animals—an important factor in the formulation of air pollution control criteria. Assuming a great expansion in nuclear power, certain radioactive waste gases could create local and regional hazards to human health and the natural environment by the beginning of the next century. Such hazards require reappraisal.

5   At a slightly lower level of priority the problems of noise and amenity, agricultural wastes and water pollution, and the effects of air pollution upon materials and some agricultural plants can be identified.

6   These might be followed by the effects of localised atmospheric

industrial pollutants upon soils, damage to land and the pollution of water courses by solid waste disposal, the selection of suitable plants for planting in areas with polluted soil and atmosphere, and the effects of odours upon local amenity. Localised problems of thermal pollution might also be included in this last group.

Fortunately much relevant research is now under way.[182, 183] Furthermore changes in technology, waste disposal methods, the value placed by society upon damage, and supplies of non-renewable resources contributing to pollution could eliminate many pollution problems but, at the same time, create new ones. Irrespective of the importance of individual pollutants, there remains a fundamental need for estimates of damage to assist the development of rational pollution control policies.

References

1   Lee, N., and Saunders, P. J. W. (1973), 'Environmental pollution: a guide to recent publications', *Brit. Book News*, March, 140–9. (This will provide a starter for those seeking useful general and special references.)

2   McLoughlin, J. (1972), *The law relating to pollution*. Manchester University Press.

3   Stoklasa, J. (1923) *Die Beschadigung der Vegetation durch Rauchgase und Fabriksexhalation*. Urban & Schwartzburg, Berlin.

4   Bell, J. N. B., and Clough, W. S. (1973), 'Depression of yield in rye grass exposed to sulphur dioxide'. *Nature, Lond., 241*, 47–9.

5   Landau, E., and Brandt, C. S. (1970), 'The use of surveys to estimate air pollution damage to agriculture'. *Environ. Res. 3*, 54–61.

6   Guderian, R., van Haut, H., and Stratmann, H. (1960) 'Problems of the recognition and evaluation of the effects of gaseous air pollutants upon vegetation', *Zeits. Pflanzenkrank, L. 67*, 257–64.

7   Moore, N. W. (1969), 'Experience with pesticides and the theory of conservation'. *Biol. Conserv. 1, 201–7.

8   Van Haut, H., and Stratmann, H. (1969), *Farbtafelatlas uber Schwefeldioxid-Wirkungen an Pflanzen*. Giradet, Essen.

9   Victor, P. A. (1973), *Economics of pollution*. Macmillan, London.

10 Pollution Research Unit (1973), *Environmental pollution: a research report*. Final report to the Science and Social Sciences Research Councils. PRU, Manchester University.

11 Lee, N., and Luker, J. A. (1971), 'An introduction to the economics of pollution'. *Economics, 9*, 19–31.

12 Lee, N. (1970–72), internal working papers. Pollution Research Unit, Manchester University.

13 Royal Commission on Environmental Pollution (1971), *First report*, Cmnd. 4585. HMSO, London.

14 Department of the Environment (1973), *The monitoring of the environment in the United Kingdom*. HMSO, London.

15 Saunders, P. J. W., and Wood, C. M. (1972), 'Sulphur dioxide in the environment: its production, dispersal and fate', in Ferry, B. W., Baddeley, S., and Hawksworth, D. C. (eds.), *Air pollution and lichens*. Athlone Press, London.

16 Royal College of Physicians (1971), *Air pollution and health*. Pitman Medical, London.

17 Gardner, M. J., Crawford, M. D., and Morris, J. N. (1969), 'Patterns of mortality in middle and early old age in the county boroughs of England and Wales'. *Brit. J. Prev. Soc. Med. 23*, 133–40, and (1970) 24, 58–60.

18 Lave, L. B., and Seskin, E. P. (1970), 'Air pollution and human health'. *Science, 169,* 723–33.

19 Programmes Analysis Unit (1972), *An economic and technical appraisal of air pollution in the United Kingdom.* HMSO, London.

20 Department of the Environment (1974), *Lead in the environment and its significance to Man.* HMSO, London.

21 Goldsmith, J. R. (1969), 'Epidemiological bases for possible air quality criteria for lead'. *J. Air Pollut. Control Assoc. 19,* 714–21.

22 Jeffries, D. T., and French, M. C. (1972), 'Lead concentrations in small mammals trapped on roadside verges and field sites'. *Environ. Pollut. 3,* 145–56.

23 Tudor, A. (1972), *The toxic metals.* Earth Island, London.

24 Wood, C. M. (1973), 'Visibility and sunshine in Greater Manchester'. *Clean Air, 3,* 15–24.

25 Derwent, R. G., and Stewart, H. N. M. (1973), 'Ozone in central London'. *Nature, Lond., 241,* 341.

26 Committee on the Problem of Noise (the Wilson committee) (1963), *Final report,* Cmnd. 2056. HMSO, London.

27 Timbers, J. A. (1965), *Traffic survey at 1300 sites.* RRL report LR206, Road Research Laboratory. Also Road Research Laboratory (1969), *Forecasts of vehicles and traffic in Great Britain.* RRL report LR288.

28 Webster, C. C. (1967), *Effects of air pollution on plants and soil.* Agricultural Research Council, London.

29 Stern, A. C. (ed.), (1968), *Air pollution* (three vols.). Academic Press, London.

30 Saunders, P. J. W. (1971), 'Modifications of the leaf surface and its environment by pollution', in Preece, T. F., and Dickinson, C. H. (eds.), *Ecology of leaf-surface microorganisms.* Academic Press, London.

31 Gilbert, O. L. (1968), 'Bryophytes as indicators of air pollution in the Tyne valley'. *New Phytol. 67,* 15–30.

32 Allcroft, R., and Burns, K. N. (1968), 'Fluorosis in cattle in England and Wales: incidence and sources'. *Fluoride, 1,* 50–3.

33 Saunders, P. J. W., and Wood, C. M. (1974), 'Plants and air pollution'. *J. Inst. Landscape Arch. 105,* February, 28–30.

34 Cowling, D. W., and Jones, L. H. P. (1970), 'A deficiency in soil sulphur supplies for perennial rye grass in England'. *Soil Sci. 110,* 346–54.

35 Purves, D. (1972), 'Consequences of trace-element contamination of soils'. *Environ. Pollut. 3,* 17–24.

36 Askew, R. R., Cook, L. M., and Bishop, J. A. (1971), 'Atmospheric pollution and melanic moths in Manchester and its environs'. *J. Appl. Ecol. 8,* 247–56.

37 Wood, C. M., Lee, N., Saunders, P. J. W., and Luker, J. A. (1973), *The geography of pollution: a study of pollution in Greater Manchester.* Manchester University Press.

38 Wood, C. M. (1970–72), internal working papers. Pollution Research Unit, Manchester University.

39   Peirson, D. H., Cawse, P. A., Salmon, L. and Cambray, R. S. (1973), 'Trace elements in the atmospheric environment'. *Nature, 241,* 252–6.
40   Klein, L. (1959–66), *River pollution* (three vols.). Butterworth, London.
41   Holden, W. S. (ed.) (1970), *Water treatment and examination.* Churchill, London.
42   Brown, V. M. (1969), 'The calculation of the acute toxicity of mixtures of poisons to rainbow trout'. *Wat. Res. 2,* 723–33.
43   Natural Environment Research Council (1972), *Research in freshwater biology.* NERC Publications, series B(2). NERC, London.
44   Department of the Environment and the Welsh Office (1971), *Report of a river pollution survey of England and Wales, 1* (1972) (updated 1972), and (1973) *2.* HMSO, London.
45   Saunders, P. J. W. (1970–72), internal working papers. Pollution Research Unit, Manchester University.
46   Natural Environment Research Council (1971), *Main report of the national angling survey, 1970.* NERC, London.
47   Vollenweider, R. A. (1968), *Scientific fundamentals of the eutrophication of lakes and flowing waters, with particular reference to nitrogen and phosphorus as factors in eutrophication.* Technical report DAS/CSI/68.27. Organisation for Economic Co-operation and Development (OECD), Paris.
48   Downing, A. L. (1970), 'Review of national research policy in eutrophication problems'. *Wat. Treat. Examin. 19,* 223–38.
49   World Health Organisation (1970), *European standards for drinking water.* WHO, Geneva.
50   Schroeder, H. A. (1966), 'Municipal drinking water and cardiovascular death rates'. *J. Amer. Med. Assoc. 195,* 81–5.
51   Toms, R. H. (1970), *The threat to inland waters from oil pollution.* Proceedings of Seminar on Water Pollution by Oil, Aviemore (May 1970), offprint.
52   Department of the Environment (1971), *Report of the working party on refuse disposal.* HMSO, London.
53   Department of the Environment (annual), *Derelict land—summary of returns.* DOE, London, and county council planning departments.
54   Tanfield, D. A. (1969), *Disposal and reclamation of colliery spoils.* National Coal Board, London.
55   Lee, N., and Saunders, P. J. W. (1972), 'Pollution as a function of population and affluence', in Cox, P. R., and Peel, J. (eds.), *Population and pollution.* Academic Press, London.
56   Food and Agricultural Organisation (1971), *Report of the FAO Technical Committee on Marine Pollution and its Effects on Living Resources.* FAO Fisheries Report No. 99. FAO, Rome.
57   O'Sullivan, J. (1972), 'Marine pollution'. *Wat. Pollut. Control* (1972), 312–32.
58   Wood, P. C. (1961), 'The production of clean shellfish (I)'. *Proc. Roy. Soc. Hlth,* 25 January, 7–11.
59   Irukayama, K. (1967), 'The pollution of Minamata Bay and Minamata

disease'. *Proc. Third Int. Conf. Wat. Pollut. Res., Munich (1966) 3*, 153–80.

60 Nelson-Smith, A. (1971), *Oil pollution and marine ecology*. Elek Science, London.

61 Food and Agricultural Organisation (1972), *Marine pollution and sea life*. Fishing News (Books) Ltd, London.

62 Royal Commission on Environmental Pollution (1972), *Third report*, Cmnd 5054. HMSO, London.

63 Porter, E. (1973), *Pollution in four industrialised estuaries: four case studies undertaken for the Royal Commission on Environmental Pollution*. HMSO, London.

64 Preston, A. (1973), 'Heavy metals in British waters'. *Nature, Lond., 242*, 95–7 (and others).

65 Agricultural Research Council (1970), *Third report of the Research Committee on Toxic Chemicals*. HMSO, London.

66 Rudd, R. L. (1965), *Pesticides in the living landscape*. Faber & Faber, London.

67 Mellanby, K. (1970), *Pesticides and pollution*. Collins, London.

68 Cooke, A. S. (1973), 'Shell thinning in avian eggs by environmental pollutants'. *Environ. Pollut. 4*, 85–152.

69 Prestt I., Jeffries, D. J., and Moore, N. W. (1970), 'Polychlorinated biphenyls in wild birds in Britain and their avian toxicity'. *Environ. Pollut 1*, 3–26.

70 Maugh, T. H. (1973), 'DDT: an unrecognised source of polychlorinated biphenyls'. *Science, N.Y., 180*, 578–9.

71 Organisation for Economic Co-operation and Development (1973), 'OECD Council takes major decision with regard to the control of certain toxic chemicals'. Press report, OECD, Paris.

72 Department of Trade and Industry (annual), *Report of the Government Chemist*. HMSO, London.

73 Ministry of Agriculture, Fisheries and Food (1972), *Survey of lead in food*. MAFF, London. (See also ref. 20.)

74 US Congress Joint Committee on Atomic Energy (1969–70), *Environmental effects of producing nuclear power* (three parts). USGPO, Washington, D.C.

75 Mitchell, N. T. (1967–72), *Radioactivity in surface and coastal waters of the British Isles*. Ministry of Agriculture, Fisheries and Food, FRL series. Fisheries Radiological Laboratories, Lowestoft.

76 Preston, A. (1970), *The UK approach to the application of ICRP standards to the controlled disposal of radioactive waste resulting from nuclear power programmes*. IAEA/UNAEC Symposium on Environmental Aspects of Nuclear Power, New York (1970), 1–7.

77 Beattie J. R., and Bryant, P. M. (1970), *Assessment of environmental hazard from reactor fission product releases*. AHSB (S) R135. HMSO, London.

78 Farmer F. R. (1967), *Siting criteria—a new approach*. UKAEA/IAEA symposium, Vienna (1967), SM 89/34. AERE, Risley.

79   Wray, E. T. (1970), *Environmental monitoring associated with discharges of radioactive waste during 1969 from UKAEA establishments.* AHSB (RP) R105. UKAEA, Harwell (and others).

80   Dobbs, C. G. (1957), 'The safety of fluorides in water'. *Brit. Dent. J. 103*, 267–74.

81   Natural Environment Research Council (1972), *Monks Wood experimental station: report for 1969–71.* Nature Conservancy, Abbots Ripton.

82   Abbott, D. C., Holmes, D. C., and Tatton, J. O'G. (1969), 'Pesticide residues in the total diet in England and Wales, 1966–67. II, Organochlorine pesticide residues in the total diet'. *J. Sci. Fd. Agric. 20*, 245–9 (and others).

83   Department of the Environment (the Central Unit on Environmental Pollution has studied the non-agricultural uses of pesticides; the results should be published in the near future).

84   Dunster, H. J., and Warner, B. F. (1970), *The disposal of noble gas fission products from the reprocessing of nuclear fuel.* AHSB (RP) R101. HMSO, London.

85   European Nuclear Energy Agency and Organisation for Economic Co-operation and Development (1971), *Radioactive waste management practices in western Europe.* OECD, Paris.

86   Gilbert, O. L. (1973), 'The effect of airborne fluorides', in Ferry, B. W., Baddeley, S., and Hawksworth, D. C. (eds.), *Air pollution and lichens.* Athlone Press, London.

87   Holdgate, M. W. (ed.) (1971), *The seabird wreck in the Irish Sea, autumn 1969.* Natural Environment Research Council publications, series C (4). NERC, London.

88   Study of Critical Environmental Problems (1971), *Man's impact on the global environment: assessments and recommendations for action.* Massachusetts Institute of Technology Press, London (and others in series).

89   Polunin, N. (ed.) (1972), *The environmental future.* Macmillan, London.

90   Bossavy, J. (1969), in *Air pollution: Proc. First Euro. Congr. Wageningen (22–27 April 1968),* 15–25. Centre for Agricultural Publishing and Documentation, Wageningen.

91   Peterson, J. T. (1969), *The climate of cities: a survey of recent literature.* National Air Pollution Control Administration, Washington, D.C.

92   Royal Ministry of Foreign Affairs and Royal Ministry of Agriculture (1971) *Air Pollution across national boundaries: the impact on the environment of sulphur in the air and precipitation.* Norstedt & Soner, Stockholm.

93   Holden, A. V., and Marsden, K. (1967), 'Organochlorine pesticides in seals and porpoises'. *Nature, Lond., 216,* 1274–6.

94   Organisation for Economic Co-operation and Development (1970), *Joint OECD/TNO conference on the occurrence and significance of pesticide residues in the environment, Helvoirt, Netherlands, 15–18 September 1969* (and later OECD publications). OECD, Paris.

95   Food and Agricultural Organisation (1971), *Joint Group of Experts on the Scientific Aspects of Marine Pollution (GESAMP): report of third session, Rome, February 1971.* GESAMP (11/19). FAO, Rome (and others).

96   Jeffries, D. J., and French, M. C. (1971), 'Hyper- and hypothyroidism in pidgeons fed with DDT: an explanation of the "thin eggshell" phenomenon'. *Environ. Pollut. 1,* 235–42.

97   Harvey, G. (1971), Wood's Hole Oceanographic Institute, press report, *Observer*, 12th December.

98   National Institute of Public Health (1971), 'Environmental mercury pollution'. *Nord. Lyg. Tidsher. Suppl. 4,* 364.

99   Shackette, H. T., Boerngen, J. G., and Turner, R. L. (1971), *Mercury in the environment—superficial materials of the coterminous United States.* Geological Survey circular 644. US Department of the Interior, Washington, D.C.

100 Department of the Environment (1971), *Mercury and water pollution.* Notes on Water Pollution, 55. Water Pollution Research Laboratory, Stevenage.

101 Jensen, J. S., and Jernelov, A. (1969), 'Biological methylation of mercury in aquatic organisms'. *Nature, Lond., 223,* 753–4.

102 Ministry of Agriculture, Fisheries and Food (1971), *Survey of mercury in food.* HMSO, London.

103 Food and Agricultural Organisation (1971), *Report of the seminar on methods of detection, measurement and monitoring of pollutants in the marine environment.* FAO, Rome, December 1970.

104 State of California (1967), *Lead in the environment and its effects on humans.* State Department of Public Health, Berkeley, California.

105 Bryce-Smith, D. (1971), 'Lead pollution—a growing hazard to the public'. *Chem. Britain, 7,* 54–6.

106 Department of the Environment (annual), *Alkali etc. Works Act.* HMSO, London.

107 Lawther, P. J., Commins, B. T., Ellison, J. McK., and Biles, B. (1972), 'Airborne lead and its uptake by inhalation', in *Proc. Conf. on Lead in the Environment,* ed. Hepple, 8–28. Institute of Petroleum, London.

108 Lane, R. A., *et al.* (1968), 'Diagnosis of inorganic lead poisoning: a statement'. *Brit. Med. J. 4,* 501–2.

109 Department of the Environment (1971), *Water pollution research, 1970.* HMSO, London.

110 Schroeder, H. A. (1965), 'Cadmium as a factor in hypertension'. *J. Chron. Dis. 18,* 647–8 (and others).

111 Lund, J. W. G. (1972), 'Eutrophication'. *Proc. R. Soc. Lond. B 180,* 371–82.

112 Lanff, G. H. (ed.) (1967), *Estuaries.* American Association for the Advancement of Science, Washington, D.C.

113 Ingham, H. R., Mason, J., and Wood, P. C. (1968), 'Distribution of toxin in molluscan shellfish following the occurrence of mussel toxicity in north-east England'. *Nature, Lond., 220,* 25–7 (and others).

114 Loehr, R. C. (1968), *Pollution implications of animal wastes—a for-*

ward orientated review. Fresh Water Pollution Control Administration, Washington, D.C.

115 Fresh Water Pollution Control Administration (1970), *National estuarine inventory*. USPGO, Washington, D.C.

116 Ricker, W. E. (1969), *Food from the sea: resources and man*. Freeman, San Francisco.

117 Gilmore, G. A., Smith, D. D., Rice, H. H., Shenton, E. H., and Moser, W. H. (1970), *Systems study of oil spill clean-up procedures, 1*. Dillingham Environmental Corporation, La Jolla, California.

118 Lloyd, R., (1972), 'Problems in determining water quality criteria for freshwater fisheries'. *Proc. R. Soc. Lond. B 189*, 439–49.

119 Alabaster, J. S. (1969), 'Evaluating risks of pesticides to fish'. *Proc. Fifth Br. Insectic. Fungic. Conf.* 370–7.

120 Thomas, M. D., and Hendricks, R. H. (1956), 'Effects of air pollution on plants', in Magill, P. L., Holder, H. R., and Ackley, C. (eds.), *Air pollution handbook*. McGraw-Hill, New York.

121 Hill, A. C. (1969), 'Air quality standards for fluoride vegetation effects'. *J. Air Pollut. Control Assoc. 19*, 331–6.

122 Dreisinger, B. R., and McGovern, P. C. (1970), 'Monitoring of atmospheric sulphur dioxide and correlating its effects on crops and forests in the Sudbury region'. *Proc. Conf. Impact of Air Pllut. on Veg., Toronto (1970)*. Offprint.

123 Sprague, J. B. (1971), 'Measurement of pollutant toxicity to fish. III. Sub-lethal effects and safe concentrations'. *Wat. Res. 5*, 245–66 (and others).

124 Jackson, S. M., and Brown, V. M. (1969), 'Effects of toxic wastes on treatment processes and water courses'. *Inst. Wat. Pollut. Control Ann. Conf., Douglas (1969)*. Conference paper 2c.

125 Coburn, R. F. (ed.) (1970), 'Biological effects of carbon monoxide'. *Ann. N.Y. Acad. Sci. 174*, 1–430.

126 Viraraghavan, T. (1971), 'Mercury pollution', *Wat. Waste Treat.*, September, 8–9.

127 Lucas, J. A. (1973), *Our polluted food*. Wright & Sons, Bristol.

128 Holden, A. V. (1965), 'Contamination of freshwater by persistent insecticides and their effects on fish'. *Ann. Appl. Biol. 55*, 322–5.

129 Eagers, R. Y. (1969), *Toxic properties of inorganic fluorine compounds*. Elsevier, London.

130 Zahn, R. (1970), 'Wirkung von Kombination unterschwellinger und toxischer Schwefeldioxid-dosen auf Planzen'. *Staub Reinhault, 30*, 162–4.

131 Brown, V. M., Shurben, D. G., and Shaw, D. (1970), 'Studies on water quality and the absence of fish from some polluted English rivers'. *Wat. Res. 4*, 362–82.

132 Brandt, C. J., and Rhoades, R. W. (1972), 'Effects of limestone dust accumulation on the composition of a forest community'. *Environ. Pollut. 3*, 217–26.

133 Howells, G. P. (1972), 'The estuary of the Hudson river, USA'. *Proc. R. Soc. Lond. B. 180*, 521–34.

134   Yapp, W. B. (1972), *Production, pollution and protection*. Wykeham Science series 19, London.

135   Katz, M. (1969), *Measurement of air pollutants: guide to selection of methods*. World Health Organisation, Rome.

136   Kenneweg, H. (1970), 'The problem of recognising and demarcating fume damage on aerial photos'. *Proc. Int. Syrup. Forest Fume Damage Experts, Essen, 1970.* Offprint.

137   Moore, N. W. (1966), 'A pesticide monitoring system with special reference to the selection of indicator species'. *J. Appl. Ecol. 3* (suppl.), 261–9.

138   Ferry, B. W., Baddeley, S., and Hawksworth, D. C. (eds.) (1973), *Air pollution and lichens*. Athlone Press, London.

139   Smith, J. E. (ed.) (1969), *Torrey Canyon*. Cambridge University Press.

140   Ayres, S. M., and Buehler, M. E. (1970), 'Effects of urban air pollution on health'. *Clin. Pharmacol. Therapeutics, 11,* 337–71.

141   Manley, P. F. J. (1971), 'Estimates of marking effect with capture–recapture sampling'. *J. Appl. Ecol. 8,* 181–9.

142   Paul, O. (1967), 'Value of prospective studies'. *Proc. R. Soc. Med. 60,* 53–6.

143   Colley, J. R. T., and Reid, D. D. (1970), 'Urban and social origins of childhood bronchitis in England and Wales'. *Brit. Med. J. 2,* 213–17.

144   International Commission for Radiological Protection (1969), *Radiosensitivity and spatial distribution of dose*. ICRP publications, 14. Permagon Press, London.

145   Sprague, J. B., Elson, P. F., and Duffy, J. R. (1971), 'Decreases in DDT residues in young salmon after forest spraying in New Brunswick'. *Environ. Pollut. 1,* 191–203.

146   Doll, R., and Hill, A. B. (1964), 'Mortality in relation to smoking: ten years' observations on British doctors'. *Brit. Med. J. 1,* 139–410 and 1460–7.

147   Zahn, R. (1961), 'Effects of sulphur dioxide on vegetation: results of experimental exposure to gas'. *Staub, 21,* 56–60.

148   Zahn, R. (1963), 'The effect of various environmental factors on the susceptibility of plants to sulphur dioxide'. *Z. Pflanzenkrank. 70,* 81–95.

149   Ball, I. R. (1967), 'The relative susceptibilities of some species of freshwater fish to poisons. II. Zinc'. *Wat. Res. 1,* 777–83.

150   Prinz, B., and Stratmann, H. (1970), 'Possibility of the analytical determination of dose-threshold value relationships'. *Staub Reinhalt 30,* 372–5.

151   Heck, W. W., and Tingey, D. T. (1970), *Models for dose–response curves of plants to several gaseous air pollutants*. Proc. Second Int. Clean Air Congress, Washington, D.C. Preprint.

152   Hayes, W. J. (1969), 'Toxicity of pesticides to man'. *Proc. R. Soc. Lond. B. 167,* 101–27.

153   Gibson, S. L. M., Mackenzie, J. C., and Goldberg, A. (1968), 'The diagnosis of industrial lead poisoning'. *Brit. J. Indust. Med. 25,* 40–51.

154   Zahn, R. (1963), 'The significance of continuous and intermittent sul-

phur dioxide action for plant reaction'. *Staub, 23,* 343–52.

155 Lambert, P. M., and Reid, D. D. (1970), 'Smoking, air pollution and lung cancer'. *Lancet, 1,* 853–7.

156 Bridges, B. (1971), 'Environmental genetic hazards—the impossible problem?' *Ecologist, 1,* 19–21.

157 Noren, K., and Westoo, G. (1967), 'Metylkvicksilver i fisk'. *Var Foda, 19,* 1–24.

158 Westoo, G., and Noren, K. (1967), 'Kvicksilver och metylkvicksilver i fisk'. *Var Foda, 19, 153*–78.

159 Gutman, H., and Serada, P. J. (1968), *Metal corrosion by the atmosphere.* American Society for Testing Materials, STP 435, Washington, D.C.

160 Hajduk, J. (1970), 'Enwirkingen von Industrie—Exhalationen auf die Strukter der Phytocoenosen'. *Geselleschaftsmorphologie,* 340–9 (and others).

161 Guderian, R. (1966), 'Reactions of agricultural fodder crops to the influence of sulphur dioxide'. *Schriftenreihe des Landesanstalt für Immissions und Bodennutzungschutz des LNW. Essen, 4,* 80–100.

162 See ref. 132 and contrast with Darley, E. F., and Middleton, J. L. (1966), 'Problems of air pollution in plant pathology. *A. Rev. Phytopath. 4,* 103–118.

163 Fresh Water Pollution Control Administration (1967), *Pollutional effects of pulp and paper mill wastes in Puget Sound.* FWPCA, Portland, Oregon.

164 Balazova, G. (1971), 'The effects of a prolonged industrial atmospheric pollution from fluorine on children'. *Medna. Lav. 62,* 202–7 (and others).

165 Hawksworth, D. L., and Rose, F. (1970), 'Qualitative scale for estimating sulphur dioxide air pollution in England and Wales using epiphytic lichens'. *Nature, Lond., 227,* 145–8.

166 Department of the Environment (1972), *Out of sight, out of mind; report of a working party on sludge disposal in Liverpool Bay, 1.* HMSO, London.

167 Fimreite, N. (1970), 'Mercury uses in Canada and their possible hazards as sources of mercury contamination'. *Environ. Pollut. 1,* 119–32.

168 Bangay, C. E. (1971), *Production and use of mercury in Canada.* Discussion paper 71–8. Policy Planning Research Service, Ottawa.

169 Pollution Research Unit (1971–72), internal working papers on *Industrial case studies.* PRU, Manchester University.

170 Strom, A. C. (1968), 'Atmospheric dispersion of stack effluents', in Stern, A. C. (ed.) *Air Pollution, 1.* Academic Press, London.

171 Fink, F. W., Buttner, F. H., and Boyd, W. K. (1971), *Technical–economic evaluation of air pollution corrosion costs on metals in the US.* Battelle Memorial Institute, Columbus, Ohio.

172 Lloyd, R., and Jordan, D. H. M. (1964), 'Predicted and observed toxicities of several sewage effluents to rainbow trout: a further study'. *J. Proc. Inst. Sew. Purif.* 167–73.

173   Dunster, H. J., Garner, R. J., Howells, H., and Wix, L. F. U. (1969), 'Environmental monitoring associated with the discharge of low-activity radioactive waste from Windscale works to the Irish Sea'. *Hlth. Phys. 10*, 353–62.
174   Preston, A. (1970), See ref. 10 and others by same author.
175   Leone, I. A., and Brennan, E. (1972), 'Sulphur nutrition as it contributes to the susceptibility of tobacco and tomato to SO_2 injury'. *Atmos. Environ. 6*, 259–66.
176   Lunn, J. E., Knoweldon, J., and Roe, J. W. (1970), Patterns of respiratory illness in Sheffield junior schoolchildren'. *Brit. J. Prev. Soc. Medic. 24*, 223–8.
177   Tubbs, C. W., and Blackwood, J. W. (1971), 'Ecological evaluation of land for planning purposes'. *Biol. Conver. 3*, 169–72.
178   Duffield, B. S., and Owen, M. L. (1970), *Leisure + countryside = ?* Department of Geography, University of Edinburgh.
179   Clawson, M. (1959), *Methods of measuring the demand for and value of outdoor recreation.* Resources for the Future Inc, Johns Hopkins Press, Baltimore.
180   International Decade of Ocean Exploration (1972), *Baseline studies of pollutants in the marine environment and research recommendations.* IDOE baseline conference, New York, 24–26 May 1972.
181   Committee (3) of the UN Conference on the Human Environment (1972), *Identification and control of pollutants of broad international significance, subject area III.* UNO, Geneva.
182   Inter-Research Council Report (1971), *Pollution research and the Research Councils;* and (1973), *Pollution research and the Research Councils revised table of research projects.* (New edition in press.) NERC, London.
183   Department of the Environment (1972), *Index of current government and government-supported research in environmental pollution in Great Britain.* DOE, London. (A new register is in preparation.)

Index